THE PICKERING MASTERS

THE COMPLETE PLAYS OF FRANCES BURNEY

Volume 2. *Tragedies*

Frances Burney. From a portrait by Edward Francesco Burney, c. 1784–5.
From the collection at Parham Park, Sussex.

THE COMPLETE PLAYS OF
FRANCES BURNEY

VOLUME 2
TRAGEDIES

Edited by
PETER SABOR

Associate Editor
STEWART J. COOKE

Routledge
Taylor & Francis Group
LONDON

First published 1995 by Pickering & Chatto (Publishers) Limited

Published 2016 by Routledge
2 Park square,Milton Park,Abingdon,Oxon OX14 4RN

Routledge is an imprint of the Tavlor & Francis Group, an informa business

British Library Cataloguing in Publication Data
Burney, Fanny
Complete Plays of Frances Burney. –
Vol. 2. – (Pickering Masters Series)
I. Title II. Sabor, Peter III. Series
822.6
Set ISBN 1 85196 073 2
ISBN 13: 978-1-138-75883-4 (Vol. 2) (hbk)

Published simultaneously in Canada by McGill-Queen's University Press

Typeset by Waveney Typesetters Norwich

CONTENTS

TRAGEDIES

List of Illustrations vi

Edwy and Elgiva (1788–95) 1

Hubert De Vere (1790–97) 91

The Siege of Pevensey (1790–91) 165

Elberta (1791–1814) 229

Appendix: *The Triumphant Toadeater* (1798), by Ralph Broome or Charlotte Burney, possibly in collaboration with Frances Burney 309

LIST OF ILLUSTRATIONS

1. Frances Burney. From a portrait by Edward Francesco Burney, c. 1784–5. *From the collection at Parham Park, Sussex.* — *Frontispiece*

2. The first page of Burney's manuscript of *Edwy and Elgiva.* *By permission of the Henry W. and Albert A. Berg Collection, the New York Public Library, Astor, Lenox, and Tilden Foundations.* — 2

3. The playbill for *Edwy and Elgiva*, performed at Drury Lane, 21 March 1795. *By permission of the Trustees of the British Library.* — 3

4. Drury Lane Theatre, 1795. From a watercolour by Edward Dayes. *By permission of the Henry E. Huntington Library and Art Gallery, San Marino, California.* — 4

Edwy and Elgiva

A Tragedy

By the Author of *Evelina* and *Cecilia*

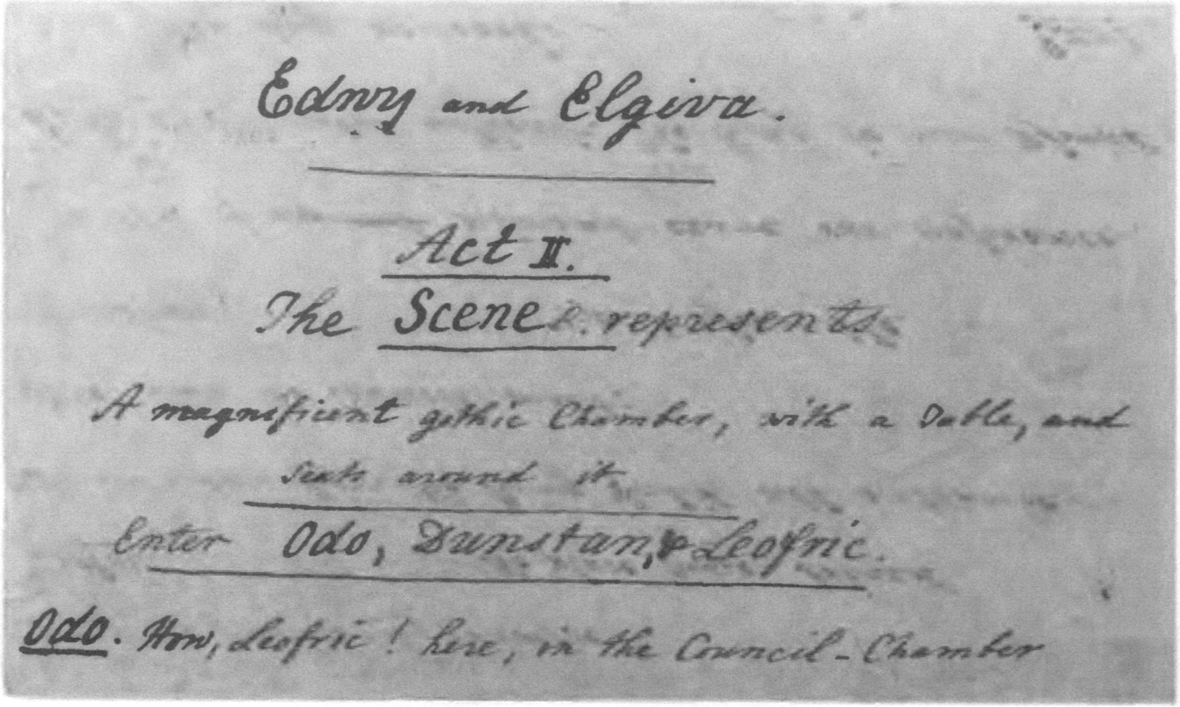
Edwy and Elgiva.

Act II.

The Scene represents

A magnificent gothic Chamber, with a Table, and Seats around it.

Enter Odo, Dunstan, & Leofric.

Odo. How, Leofric! here, in the Council-Chamber

The first page of Burney's manuscript of *Edwy and Elgiva*. *By permission of the Henry W. and Albert A. Berg Collection, the New York Public Library, Astor, Lenox, and Tilden Foundations.*

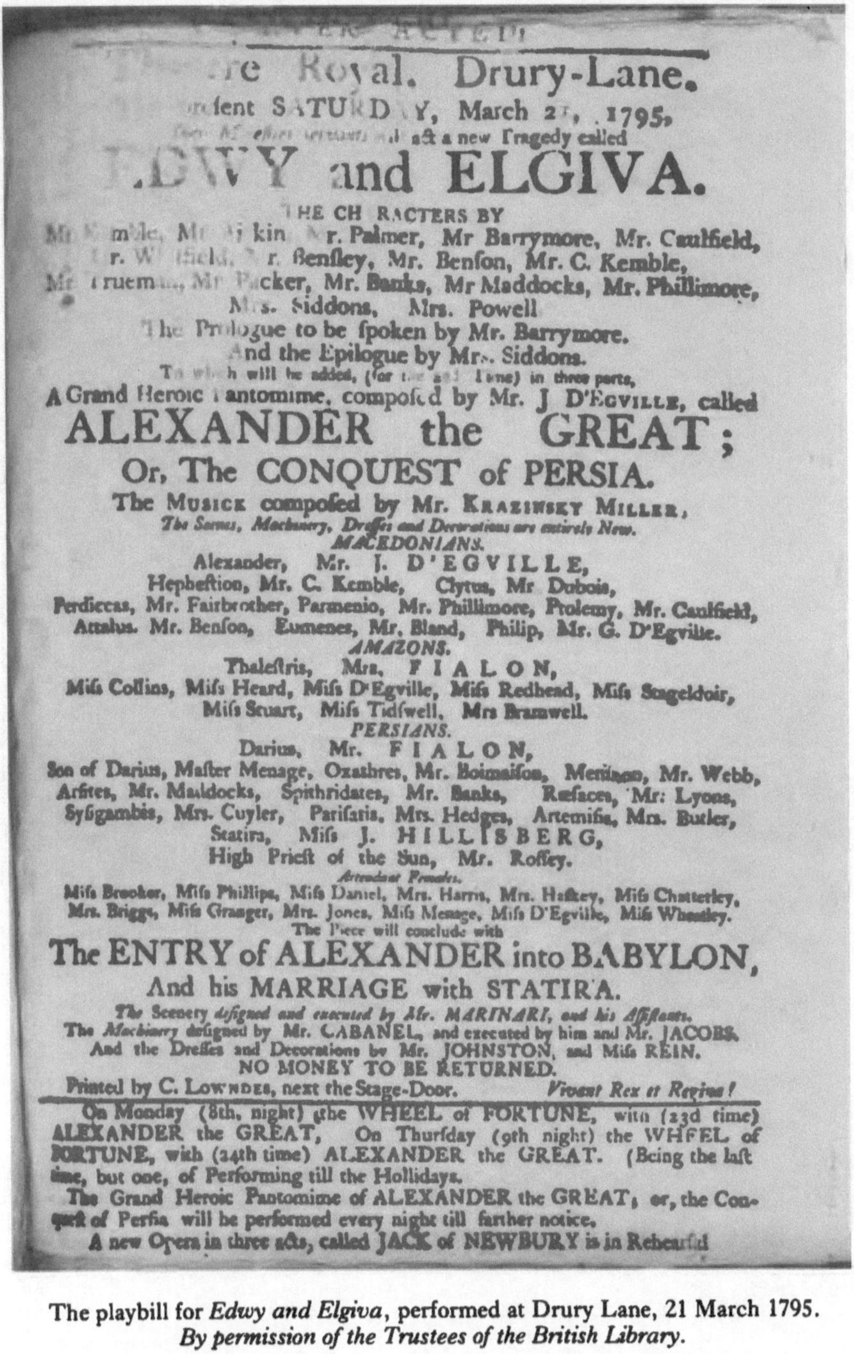

NEVER ACTED.

Theatre Royal, Drury-Lane.

This present SATURDAY, March 21, 1795,

Their Majesties Servants will act a new Tragedy called

EDWY and ELGIVA.

THE CHARACTERS BY

Mr. Kemble, Mr. Aickin, Mr. Palmer, Mr Barrymore, Mr. Caulfield,
Mr. Whitfield, Mr. Bensley, Mr. Benson, Mr. C. Kemble,
Mr. Trueman, Mr. Packer, Mr. Banks, Mr Maddocks, Mr. Phillimore,
Mrs. Siddons, Mrs. Powell

The Prologue to be spoken by Mr. Barrymore.
And the Epilogue by Mrs. Siddons.

To which will be added, (for the 22d Time) in three parts,
A Grand Heroic Pantomime, composed by Mr. J D'EGVILLE, called

ALEXANDER the GREAT;

Or, The CONQUEST of PERSIA.

The MUSICK composed by Mr. KRAZINSKY MILLER,
The Scenes, Machinery, Dresses and Decorations are entirely New.

MACEDONIANS.

Alexander, Mr. J. D'EGVILLE,
Hephestion, Mr. C. Kemble, Clytus, Mr Dubois,
Perdiccas, Mr. Fairbrother, Parmenio, Mr. Phillimore, Ptolemy, Mr. Caulfield,
Attalus. Mr. Benson, Eumenes, Mr. Bland, Philip, Mr. G. D'Egville.

AMAZONS.

Thalestris, Mrs. FIALON,
Miss Collins, Miss Heard, Miss D'Egville, Miss Redhead, Miss Stageldoir,
Miss Stuart, Miss Tidswell, Mrs Bramwell.

PERSIANS.

Darius, Mr. FIALON,
Son of Darius, Master Menage, Oxathres, Mr. Boimaison, Menippo, Mr. Webb,
Arsites, Mr. Maddocks, Spithridates, Mr. Banks, Reesaces, Mr. Lyons,
Sysigambis, Mrs. Cuyler, Parisatis, Mrs. Hedges, Artemisia, Mrs. Butler,
Statira, Miss J. HILLISBERG,
High Priest of the Sun, Mr. Roffey.

Attendant Females.

Miss Brooker, Miss Phillips, Miss Daniel, Mrs. Harris, Mrs. Haskey, Miss Chatterley,
Mrs. Briggs, Miss Granger, Mrs. Jones, Miss Menage, Miss D'Egville, Miss Wheatley.

The Piece will conclude with

The ENTRY of ALEXANDER into BABYLON,

And his MARRIAGE with STATIRA.

The Scenery designed and executed by Mr. MARINARI, and his Assistants.
The *Machinery* designed by Mr. CABANEL, and executed by him and Mr. JACOBS.
And the Dresses and Decorations by Mr. JOHNSTON, and Miss REIN.
NO MONEY TO BE RETURNED.

Printed by C. LOWNDES, next the Stage-Door. *Vivant Rex et Regina!*

On Monday (8th, night) the WHEEL of FORTUNE, with (23d time) ALEXANDER the GREAT, On Thursday (9th night) the WHEEL of FORTUNE, with (24th time) ALEXANDER the GREAT. (Being the last time, but one, of Performing till the Hollidays.

The Grand Heroic Pantomime of ALEXANDER the GREAT; or, the Conquest of Persia will be performed every night till farther notice.

A new Opera in three acts, called JACK of NEWBURY is in Rehearsal

The playbill for *Edwy and Elgiva*, performed at Drury Lane, 21 March 1795.
By permission of the Trustees of the British Library.

Drury Lane Theatre, 1795. From a watercolour by Edward Dayes. *By permission of the Henry E. Huntington Library and Art Gallery, San Marino, California.*

EDWY AND ELGIVA

There are three surviving manuscripts of *Edwy and Elgiva*. The first, in the neat hand of a professional copyist employed by the Drury Lane Theatre, is a volume of 81 pages. This copy was submitted for license by John Philip Kemble, the actor-manager of Drury Lane, to the Lord Chamberlain, James Cecil, Marquis of Salisbury, on 9 March 1795. In an accompanying note to the Lord Chamberlain, Kemble states that 'This Tragedy in five Acts, called Edwy and Elgiva, is designed, with the Permission of the Marquis of Salisbury, for Representation at the Theatre Royal in Drury Lane'. The play was inspected by John Larpent, Examiner of Plays from 1778 to 1824, and approved without alterations being required; no Examiner's marks were made on the manuscript. This copy is number 1058 in the Larpent Collection at the Huntington Library, Los Angeles, which in 1917 acquired the several thousand play manuscripts examined by Larpent.[a]

The Larpent manuscript of *Edwy and Elgiva* provides a text of the play similar to that performed on stage, although changes could have been made by the theatre company shortly before the performance on 21 March 1795. Unlike the other copies, the Larpent manuscript contains no scene divisions, and its stage directions are less elaborate; the setting for Act I, for example, is described simply as 'a Gothic Chamber', rather than 'A Magnificent gothic Chamber, with a Table, and seats around it'. It contains a prologue by Burney's brother Charles and an epilogue written by herself, both of which were included in the stage production, the prologue spoken by William Barrymore and the epilogue by Sarah Siddons.[b]

The second extant copy, in the small clear hand of Alexandre d'Arblay, is a volume of 67 pages, rebound in 1993. It includes the prologue but not the epilogue contained in the Larpent manuscript; both copies also provide the names of the actors in the Drury Lane production. It is at Emmanuel College, Cambridge, which received it as a gift from its librarian, Evelyn

[a] Burney's editors and biographers have overlooked this acting copy of *Edwy and Elgiva*. It is listed in Dougald MacMillan's *Catalogue of the Larpent Plays in the Huntington Library* (San Marino, California: Huntington Library, 1939), p. 176.

[b] The authorship of the prologue, previously thought to have been by Frances Burney, and of the epilogue are revealed in an unpublished letter from Charles Burney to his sister of 16 March 1795, in the Berg Collection. Reviewers of the play were highly critical of the lengthy prologue, described in *The Times* as 'a tedious descant on the Three Ages of Religion', which 'seemed almost an *age* in the delivery', and in *The True Briton* as 'being nearly as long as the ordinary Act of a Play' (23 March 1795).

Shuckburgh, in the 1880s.[a] The copy was made by d'Arblay at the request of Burney's brother Charles, her intermediary in dealing with the Drury Lane company. D'Arblay agreed to Charles's request in an unpublished letter of 12 January 1795[b] and transcribed the text shortly thereafter. This copy, as his letter indicates, was made from what was then the only manuscript of the play, Burney's holograph, which was being used by the Drury Lane company. It was sent to d'Arblay, whose transcription was then used by the Drury Lane scribe as the basis for the Larpent copy. Subsequently, d'Arblay made many detailed suggestions for revisions to the text, which he inserted between the lines of his original transcription, on blank pages interleaved between the written pages, and even on extra sheets attached to these blank pages. His copy was printed (with numerous errors) in Miriam J. Benkovitz's edition of *Edwy and Elgiva* (1957). She also transcribes d'Arblay's proposed revisions in an Appendix to her edition.[c]

The third copy, in Burney's elegant hand, is in five packets, one for each act, totalling 278 pages. It contains neither the prologue and epilogue, nor the names of the actors given in the other copies. This holograph is in the Berg Collection. In her edition of *Edwy and Elgiva*, Benkovitz describes the holograph as 'earlier' than d'Arblay's copy, but this is only partly accurate.[d] It is true that this manuscript existed before either d'Arblay's or the Larpent copies were made, and that it contains passages deleted by Burney which do not appear in those copies. But the holograph is also later than d'Arblay's or the Larpent copy, in that it contains Burney's final revisions. A few of the revisions suggested by d'Arblay appear in the holograph, but the majority do not. Conversely, some but not all of Burney's revisions to the text were incorporated by d'Arblay in his copy.

The holograph is accompanied by 58 pages of supplementary material, partly in Burney's hand, partly in that of d'Arblay, and partly in an unidentified hand. This material includes a preliminary draft and the final version of Charles Burney's prologue; a draft of what seems to be a dedication, headed merely *Edwy and Elgiva*; scattered comments on the play, including remarks on the historical background; many suggestions for further revisions; poetic renditions developing parts of the play, etc.[e] There is also a transcription by d'Arblay of the revisions made by Burney in the holograph, set out in the form of corrections to each character's part; these corrections could thus readily be incorporated in the actors' individual copies were the play to be staged again.

[a] Shuckburgh used this copy as the basis for his article, 'Madame d'Arblay', *Macmillan's Magazine*, February 1890, 291–8. I have not discovered how d'Arblay's copy came into his possession.

[b] Osborn Collection, Yale University Library.

[c] See *Edwy and Elgiva*, ed. Benkovitz (New York: Shoe String, 1957), pp. 83–93.

[d] Benkovitz, p. vi. Doody likewise terms Burney's copy an 'earlier manuscript' (*Frances Burney*, p. 409, n. 89).

[e] Part of this material, an ode entitled 'El.'s madness after the Divorce', is transcribed by Doody, *Frances Burney*, pp. 195–6.

The holograph of *Edwy and Elgiva* is used as the copy-text for the present edition. The numerous revisions made by Burney in this copy are recorded in the textual notes. The obliterated passages can be read only with considerable difficulty; a few, as noted, are entirely illegible. The dedication, in Burney's hand, is transcribed from the single copy in the supplementary material to Burney's holograph. The prologue is transcribed from the final version accompanying the holograph; the epilogue is transcribed from the only existing version, the Larpent copy.

Burney began writing *Edwy and Elgiva* at court in 1788, when George III's illness was reaching its height. In a journal entry for 19 October she wrote: 'I have just begun a tragedy . . . . it may while away the tediousness of this unsettled, unoccupied, unpleasant period' (*Diary and Letters*, IV, 118).[a] Four days later she was still seized with a 'composition fit' which 'fills up all my vacancies', but by 9 November the King's illness made it impossible for her to pursue her 'melancholy resource': 'misery so actual, living, and present, was knit too closely around me to allow my depressed imagination to fancy any woe beyond what my heart felt' (IV, 119, 155). In April 1790, on Easter Sunday, she returned to her 'long-forgotten tragedy': 'I found nothing but unconnected speeches, and hints, and ideas, though enough in quantity, perhaps, for a whole play. I have now begun planning and methodising, and have written three or four regular scenes' (IV, 362). A few days later she was still at work 'occasionally' on the play, which 'does not much enliven, but it soothes me' (IV, 365). And in mid-August 1790, Burney noted with pleasure that the first draft was complete. 'What species of a composition it may prove she is very unable to tell; she only knows it was an almost spontaneous work, and soothed the melancholy of imagination for a while, though afterwards it impressed it with a secret sensation of horror'. This draft, which Burney described as 'in its first rough state, and legible only to herself' (IV, 413), must predate the highly legible holograph in the Berg Collection. By October 1791, when she was at work on *The Siege of Pevensey*, both *Edwy and Elgiva* and *Hubert De Vere* remained in draft form, 'but having only to connect, combine, contract, & finish', she 'will not leave them undone' (*Journals and Letters*, I, 74).

Burney had revised *Edwy and Elgiva* before July 1794, when she wrote to her sister Charlotte about Charles's plans, 'a little Secret transaction', to have one of her tragedies produced at Drury Lane (III, 69). A year earlier, in July 1793, Kemble had accepted *Hubert De Vere* for production (see below, p. 94); Burney, however, withdrew it in favour of *Edwy and Elgiva* 'because it was more dramatic' (III, 258). When Kemble was shown *Edwy and Elgiva*, he 'instantly & warmly pronounced for its acceptance', and on 18 December, just three days after the birth of her son Alexander, Charles informed his sister that the proprietor of the Drury

[a] Seven years later, in a letter of 15 April 1795, Burney claimed that she had begun the play in October or November 1786, but her recollection may have been faulty (*Journals and Letters*, III, 97–8).

Lane theatre, Richard Brinsley Sheridan, whose consent was also required, wished to see the play immediately (III, 98). On 5 January 1795, Charles read the play, now accepted for production, to the assembled cast. Hester Piozzi, apparently drawing on an account from Sarah Siddons, describes this inauspicious occasion: 'when reading the Play in the Green Room before exhibition – the Actors dropt silently off, one by one and left him *all alone*'.[a]

A week later, in his letter to Charles of 12 January 1795, d'Arblay wrote of Burney's desire to give the play 'a new theatrical *perfection*', including 'any abridgement for the representation' deemed necessary by Kemble. She hoped that Kemble would provide 'a mark'd copy of the passages and parts' that could best be deleted; this seems not to have been done. Burney's plans to make 'divers corrections & alterations' to the play were in any case frustrated by her absorption in her new baby and by ill health (III, 98).

After its initial reading by Charles Burney on 5 January, no rehearsal of *Edwy and Elgiva* took place until 4 March, an unusually long and inexplicable two-month delay.[b] Subsequent rehearsals took place at regular intervals in March, although one for March 12 was 'Dismiss'd by order of Mr. Kemble'. The final rehearsal, the ninth in all, took place, as was customary, on the morning of the premiere on 21 March.[c]

Beside the names of the characters in both the Larpent and d'Arblay's copies of the play, the names of the actors are listed. D'Arblay's record, the fuller of the two, lists John Philip Kemble (1757–1823) as Edwy, James Aickin (c. 1735–1803) as Odo, John Palmer (1744–1798) as Aldhelm, William Barrymore (1759–1830) as Cerdic, Thomas Caulfield (1766–1815) as Ernulf, John Whitfield (1752–1814) as Redwald, Robert Bensley (1742–1817) as Dunstan, Robert Benson (1765–1796) as Orgar, Charles Kemble (1775–1854) as Sigebert, Mr Trueman (fl. 1793–1817) as Kenred, John Hayman Packer (1730–1806) as Leofric, Walter Maddocks (d. 1823) as the Peasant, John Phillimore (d. 1799) as the First Ruffian, Benjamin Evans (d. 1820) as the Second Ruffian, Sarah Siddons (1755–1831) as Elgiva, and Jane Powell (c. 1761–1831) as Eltruda.[d] Although the cast included the two leading actors of the day, John Philip Kemble and Sarah Siddons, and others with fine reputations, such as Kemble's brother Charles and Jane Powell, many reviewers of the performance criticized the lesser actors for forgetting their parts. In a letter of 15 April, Burney, like the reviewers, spared Siddons, John Philip Kemble, and Bensley from her censure, but claimed that the others 'were cruelly imperfect, and made blunders I blush to have pass for mine'. Palmer, in particular, 'had *but 2 lines* of his part by Heart' (*Journals and Letters*, III, 99–100). Palmer was

[a] *Thraliana*, II, 916.

[b] Hogan speculates that Charles Burney's reading 'perhaps induced the theatre managers to request Madame d'Arblay to make it somewhat more suitable for representation'. See *The London Stage*, V, cxlix, n. 201.

[c] See the notes on William Powell's set of playbills, vol. 4, in the British Library.

[d] Information on the actors is taken from *A Biographical Dictionary of Actors*.

indeed notorious in his later years for forgetting his lines, and Barrymore too 'was not regarded as a good study'; in April 1795, a month after his performance in *Edwy and Elgiva*, he had to read the prologue to Edward Jerningham's *The Welch Heiress* from the prompter's copy (*Biographical Dictionary*, I, 357, 358).[a]

In a letter to her father of 13 May 1795, Burney declared that she was 'not at all without thoughts of a future revise' of the play on stage (*Journals and Letters*, III, 107). Palmer, she specified, would have to withdraw from the cast. She again refers to the possibility of further performances in a letter to her brother of 10 June 1795, and the revisions that she and d'Arblay wrote in their respective copies of the play were probably made at that time. Some of the scraps now boxed with the manuscript in the Berg Collection seem also to have been written after the performance. One, for example, suggests a revision to make the opening of Act II more dramatic. The act

> might begin with a Banquet scene – drums & trumpets. Edwy & the court entering as from the coronation. After taking their seats, a speech of welcome from the King. During the banquet, something sarcastic between Dunstan and the King or his friends, might create altercation and afterwards sullenness. which wd. make Edwy's retiring the more natural.

Despite Burney's continued interest in the play, however, it was not performed again. And plans for publishing the play shortly after its performance were opposed by Burney's father, who in a letter of 2 April 1795 declared: 'My earnestness for *not* printing was brought on by the general opinion of others who wish you well, & of whose judgment I think highly' (III, 96, n. 1).

[a] For examples of Palmer's lapses in other parts, see *The London Stage*, V, cxvi–cxviii; and Isobel Grundy, 'Sarah Gardner: "Such Trumpery" or "A Lustre to Her Sex"?', *Tulsa Studies in Women's Literature*, 7, no. 1 (1988), 11.

PERSONS OF THE DRAMA

EDWY, King of England.
ODO, Archbishop of Canterbury.
ALDHELM, Bishop of Winchester.
CERDIC, Earl of Mercia.
ERNULF, Earl of Northumberland.
REDWALD, Earl of Devon.
DUNSTAN, Abbot of Glastonbury.
ORGAR, SIGEBERT, KENRED, } Officers belonging to the King.
LEOFRIC,
A PEASANT.

ELGIVA.
ELTRUDA.

LORDS. SOLDIERS. ATTENDANTS. RUFFIANS.

Persons of the Drama: Dunstan and Odo, as well as Edwy and Elgiva, figure prominently in Hume's *History of England*, the principal source for Burney's play. Mercia and Northumberland are two of the seven Saxon kingdoms that formed the Heptarchy; see note to *Elberta*, below, p. 234.

DEDICATION

The woes of suffering Virtue to relate;
To speak the dire effects of adverse Fate;
To raise soft pity in each gen'rous breast
For hapless innocence by vice oppress'd;
To paint the mis'ries which that vice persue
To prove its joys how poor, how frail, how few;
The blessings far superior to shew
That follow Goodness, tho' they follow slow;
In Youth a noble ardour to create
The virtuous Deeds they hear and emulate;
Be such the aim of all who dare aspire
To catch the smallest spark of wild poetic fire.
  O Truth, to Honour's Path the noble Guide
Perspicuous, sacred, pure and dignified,
Simple and unadorned; – thyself a Gem
Unaided and alone Fortune's tyde can stem;
Above evasive art, or mean design,
Of all Man's attributes nearest divine;
Unsullied, uncorrupt; by all confest
The fairest Inmate of the virtuous Breast.
Inspired by which most sacred influence
Mean tho' I am, I cannot give offence.
'Tis true I may lament my wish denied
The simple to instruct; the weak to guide;
T'enlarge th'ideas with observations wise
With wit to charm; with genius to surprise.
T'expand the youthful mind with knowledge bright,
A task of labour mingled with delight,
For oh what transport must that Bosom find
Whose counsels and whose cares have serv'd mankind.

Dedication: an editorial insertion. The only surviving copy of this piece, in the supplementary material to Burney's copy of the play, is entitled simply 'Edwy and Elgiva'. It is, apparently, a dedication of the work to 'Truth'.

---

8 *slow*: two lines are deleted, the first of which is illegible; the second reads 'The Good feel courage – and the bad feel fear'.

16 *Fortune's*: an insertion, replacing '‹Fate's›'.

21 *which*: an insertion, replacing 'thy'.

25 *T'enlarge*: the ms. reads 'Th'enlarge'.

Not to my feeble pow'r such task is giv'n
Reserved for the choice favourites of Heav'n.
For me, contented in my humble state,
Not worthy praise, yet not incurring hate,
Unnoticed I shall go the common Round,
Nor yet for virtue, or for vice renown'd.
Yet still, O sacred Truth, if thou art near
Hope is more flattering, than depressing fear;
O then this humble offering receive,
The only preference I have pow'r to give.

PROLOGUE

In those dark ages when the dread command
Of Druid Superstition rul'd the land
By heaven inspir'd some holy Seer imprest
The Christian doctrines on a Britons breast
Idolatry retreats and tenets pure
Seize on the mind undecked by outward lure
To catch the Sense no ceremonious Show
No grandeur such as later times bestow
The fragrant Censer and the embroider'd vest
The pealing organ and the altar drest
The temples fretted roof – but humbler prayer
Aim'd at the heart and found an echo there.
Yet short liv'd this Religion's golden reign:
Of heresies succeed a motley train
And primitive simplicity decays
Obscur'd by Mystery's gloom and Errors Maze
Then mark the progress a disastrous hour
In Britain fixes firm the papal pow'r
When Schismatics arise To Rome the appeal
Of blind submission and mistaken zeal
There oft paid tributes every grace secure
And vows monastic heaven itself ensure.
A pilgrim's toil the wayworn Sinner saves
While Rome more closely strains her abject slaves
When Princes Priests and Laymen jointly own
His word Supreme who fills the Pontiff's throne.
Such as our Drama's Scenes this night relate
Once Britons was your Country's hapless state
Your monarch's hopes of heaven his love his crown
Hung on the imperious Pontiffs smile or frown

Prologue: Five different versions of the prologue, by Burney's brother Charles, exist: the Larpent copy, d'Arblay's copy, two versions in the supplementary material to Burney's copy, and the original in a letter from Charles to Burney of 16 March 1795. The present text is taken from the revised version in Burney's copy. The prologue was spoken by William Barrymore, who played Cerdic.

17–18 *disastrous hour ... papal pow'r*: in his *History of England*, Hume recounts Pope Gregory's successful project for 'converting the British Saxons' (I, 28) at the end of the sixth century.

19 *Schismatics*: those who promote or countenance schism in the Church.

His queen though virtue pointed Beauty's charms
Torn by fanatic edicts from his arms
What horrors follow These shall Rome bewail
Rome proudly triumphs as she hears the tale.
Such was Religions Silver Age. Her next
Of Brass by feuds and tenets strange perplext
Till Persecutions unrelenting rage
Distains with massacres her Iron Age.
What time the Pontiff's thirst of Blood displays
The streaming scaffold and the faggots blaze
Till this disgraceful bondage Henry broke
And bade his Subjects spurn the galling yoke
Your great Forefathers hail'd the glorious day
View'd Romes stern usurpation die away
And with exulting feeling's dar'd Presage
That when their annals reach'd a future age
Their sons the dear bought blessings would secure
With Church and State from innovation pure.
Such were their hopes. O Spirit of our Sires
Breathe into every Soul a Britons fires
Give us the value of those joys to know
Which rights bequeath'd and holy truths bestow
Teach us to love our Country and her Laws
To glow united in her sacred cause
And boast with swelling hearts and loud acclaim
Our Faiths Defender and our King the same.

37–41 *Persecutions* ... *Henry broke*: in 1534, the Act of Supremacy styled Henry VIII Head of the Church of England, an event also celebrated in the last line of the Prologue. The flattering portrait of Henry as foe to papal tyranny and cruelty follows that in Hume's *History*.

ACT I

Scene i

The Scene represents
A Magnificent gothic Chamber, with a Table, and seats around it.

Enter ODO, DUNSTAN, *and* LEOFRIC.

ODO. How, Leofric! here, in the Council-Chamber
May we, unheard, converse?
LEOFRIC. No place more sacred.
Strict orders from the king forbid all entrance
Save when the Council meet.
DUNSTAN. Ere it assemble
Matters of moment crave our conference.
LEOFRIC. I'll watch without. The king is now abroad;
None other dare approach.

Exit.

Scene ii

ODO. DUNSTAN.

DUNSTAN. Holy Archbishop
Learn now this most unhallow'd mystery,
Nor wonder here no step unbidden treads.
I know thee all suspicious of report
And therefore to this full conviction bring thee:
Lo! where, beneath yon fret-work, opes a passage
To Elgiva's apartment. There the Princess,
Courting black ruin through seduction's wiles,
Shameless, allures the rash young King: or thence
Hither she comes, to list to lawless vows,
Of impious love, and fabricate vain schemes
Of impious union.

I i 5 *of*: 'deepest' is deleted.
I ii 2 *now*: 'from me' is deleted.
I ii 10 *vows*: an illegible line is deleted.
I ii 12 *impious*: an insertion, replacing 'legal'.

ODO. Most prophane design!
Within the limits of affinity
By canon-Law proscrib'd, would Edwy dare –
DUNSTAN. Odo, what dare he not? Presumptuous Youth!
Does he not brave even Me, whose Will was Law
In Edred's reign? Proud Boy! think he with Edred
Died Dunstan's power? No! it yet lives in Edgar.
ODO. A younger Brother, and so mere a stripling,
O what avail is Edgar's patronage?
DUNSTAN. To time leave that. Unite we now to strengthen
The general interest of the Hierarchy.
The Convents, groaning with unrighteous rule,
Demand the expulsion of the seculars,
And loudly claim a purifying clearance.
ODO. Dunstan, most just; And at this morning's Council
We may discuss –
DUNSTAN. Discuss? no, no; decide!
Why take the lingering mode of argument,
Colour'd, chamelion-like, with every tint
Of soft persuasion, or alarming menace
To turn from side to side wayward opinion?
And hold the temper of the mind suspended?
Teaching the very populace to contend,
And weigh in Reason's scales Sedition's rights?
'Twere folly to employ such dangerous powers.
Brief be we, then, of speech, and quick to action.
ODO. Enough; the ensuing Council fixes all.

I ii 14 *canon-Law proscrib'd*: Elgiva was a second or third cousin of Edwy; ecclesiastical, though not civil, law prohibited a marriage between such partners. In a working note on the play, Burney wrote that Odo could denounce marriages between cousins, exclaiming 'that the Law against such weddings was among the Laws commanded by St Gregory and enacted by his [i.e. Odo's] great predecessor, Augustine the 1st who was consecrated archbishop of Canterbury' (Berg Collection, together with *Elberta* scraps). In the play itself, however, no mention is made of either Gregory, Pope from 590 to 604, or of Augustine, first Archbishop of Canterbury, Gregory's representative in England from 597 to 604.
I ii 17 *Edred's reign*: Edwy's predecessor, his uncle Edred, who died in 955, was a strong supporter of Dunstan, appointing him as both his spiritual advisor and head of the treasury.
I ii 22 *the Hierarchy*: the Papacy.
I ii 23–25 *The Convents ... clearance*: expelling secular priests from the convents was part of the programme of ecclesiastical reform advocated by Dunstan and his followers, who wished to enforce celibacy among the clergy and who denounced the prevalence of libertinism and luxury. Edwy was firmly opposed to the monks' policies.
I ii 34 *Sedition's rights*: the reviewer for the *True Briton*, misquoting this phrase, complained that '"seditious rights" ... we conceive to be nonsense – for sedition can have no rights – nor can the assertion of legal rights be termed sedition' (23 March 1795).

I ii 26 *just*: an insertion, replacing 'right'.
I ii 28 *Why*: 'should we' is deleted.

DUNSTAN. Be strenuous, and secure. The seculars
Have Aldhelm for a Friend; Aldhelm has Edwy.
ODO. I marvel Aldhelm should betray our Cause,
Prelate himself, and bound to be its guardian.
DUNSTAN. Odo, I rank him as our mightiest foe:
Tutor and favourite of the impetuous Edwy
Who vests him with my lost authority,
His love he favours, My Reform opposes,
And means to elude the approaching Coronation
Till Elgiva may wear a Diadem.
Let us then firm combine –

Scene iii

ODO. DUNSTAN. LEOFRIC.

LEOFRIC. Haste, haste away!
Despatch! – the king returns. (*apart to* DUNCAN) 'Tis mine to bear
The tydings to the Princess, who, perchance,
May instant hither fly.
DUNSTAN. We will begone.
LEOFRIC. Turn to the left and wait. When he is enter'd
I will conduct you thence with speed and safety.
But haste, or I am undone – for oft the king
Precludes my signal.

LEOFRIC *passes through a private door, concealed by gothic wood-work.*

Scene iv

ODO. DUNSTAN.

DUNSTAN. That good Leofric
Informed me of this wily business.
His Conscience is my own. Soft! to the left!

ODO *and* DUNSTAN *walk silently out.* LEOFRIC *re-opens the private Door, at which* ELGIVA *presently appears. He then opens the state Door in the centre, which he closes after the entrance of* EDWY, *shutting himself out.*

I ii 38 *Be strenuous ... seculars*: an insertion, replacing 'It hangs on our exertions. The bold seculars'.
I ii 47 *Elgiva*: 'as Queen' is deleted.
I iii 8 *Precludes*: an insertion, replacing 'Anticipates'.
I iv 2 *Informed me of*: an insertion, replacing 'Instructed me in all'.

Scene v

EDWY. ELGIVA. *Meeting.*

EDWY. My Elgiva!
ELGIVA. My Lord!
EDWY. My loveliest Elgiva!
O gentlest, best! – where hast thou been so long? –
Enter undaunted.
ELGIVA. I fear'd to be too soon,
And chid the willing steps that urg'd me forward; –
But they refused to heed me!
EDWY. How I thank them!
They are joy's Herald's to my faithful Breast.
If I but hear them, ev'n in Fancy's Ear,
Hope's quick pulsation vibrates through my frame,
And every nerve is strung to happiness:
Soft Expectation paints thy near approach,
And with Thee brings, Life's dearest consolation,
The softening balm of Heart-felt simpathy.
ELGIVA. If to your partial Ear, my much lov'd Lord,
Such sounds anticipate some little pleasure,
Think what to mine must Edwy's footsteps bring?
Peace cloath'd in gladness, Gratitude, in joy,
Duty array'd with smiles, and gay obedience! –
But whither run I on? –
EDWY. Why check such sweetness?
My Bride! my Wife!
ELGIVA. Ah! Venture not such sounds! –
EDWY. They will be soon familiar o'er the Land:
Aldhelm hath even now conven'd a council
To publish to the World my fix'd resolve.
ELGIVA. Knows Aldhelm yet himself my height of Honour?
EDWY. He knows thy virtues – can he doubt their sway? –
But for the rest, Eltruda and the Priest
Alone are in our secret. Aldhelm's task
Is simply to propose, not own our nuptials;
And to procrastinate My coronation
Till My sweet Partner share in all it's splendour.
For this his word inviolate is pledg'd.
Why looks my Elgiv pensive?

I v 30 *word inviolate is*: an insertion, replacing 'noble word is fully'. Before the insertion, 'pious' is deleted.

ELGIVA. Dear my Lord,
Is not the Council summon'd to this Chamber?
EDWY. Be not alarm'd; Aldhelm will first approach
To take My final orders. Why so fearful?
ELGIVA. Should Aldhelm – should the Council – should the Nation
Deem me unworthy this bright eminence –
EDWY. Know thyself better, Elgiv, and know Me:
Each claim, acquir'd and natural, is thine,
While to assert them is My easy task.
ELGIVA. Yet – when my mind revolves the Canon-Law
Prohibiting a union such as ours. –
EDWY. Revolve it not! 'tis but a prejudice,
An ord'nance of the Pope, and not from Heaven!
'Tis true, thou wert my kin'swoman; be so still,
Join'd to a softer, kinder, dearer name.
ELGIVA. I have no fear, my Lord, if you have none;
I have no dread, if you are free from doubt.
My Honour rests on your's; my Happiness
My Faith, my Trust, all own no other Guardian.
EDWY. Sacred deposits! with my life I'll shield them!
Rest then secure; I had not sought thee, Elgiv,
With all thy charms, and all their ardent power
O'er my bewitch'd, my fascinated Heart,
Could any guilt have sullied thy fair fame
In listening to my vows.
ELGIVA. My dearest Lord!
'Twas but a transient fear, and now 'tis vanish'd.
EDWY. I had not woo'd to lose thee, Elgiva;
And, let one blemish strike my watchful mind,
My Love, all founded on thy perfect virtue,
Would die, in missing its original food.
ELGIVA. O Edwy! can I merit Love like this? –
EDWY. What was it, Elgiv, that I priz'd in Thee?
O not alone the beauty of thy Form,
But the fair artless Soul that Form displays.
I see it in Thee all! thy Eye, thy air,
Thy smile, thy serious look, all, all proclaim
The spotless texture of the inward fabric
In perfect harmony with this fair promise.
To me 'tis all transparent! 'tis a lustre

I v 34 *fearful*: an insertion, replacing 'anxious'.
I v 54 *Could any guilt have sullied*: an insertion, replacing 'If any guilt could sully'.
I v 59 *And, let ... mind*: an insertion, replacing 'For, could I see in thee one spot, one blemish'.
I v 67 *spotless*: an insertion, replacing 'similar'.
I v 68 *perfect*: an insertion, replacing 'softest'.

So blending spirit with mortality
That while I hold thy Hand, and call Thee Mine,
I almost see the Wings which waft thee upward.
ELGIVA. O my lov'd Lord –

Scene vi

EDWY. ELGIVA. LEOFRIC.

LEOFRIC. The Bishop, Sir, –
EDWY. Aldhelm of Winchester?

LEOFRIC *bows.*

ELGIVA. Let me retire.

EDWY *hands her to the private Door, and returns.*

Scene vii

EDWY. LEOFRIC.

EDWY. Bring in the Bishop.

LEOFRIC *goes out.*

Scene viii

EDWY

EDWY. Sweet, softest excellence!
Heaven guard her from distress and agitation! –
I dare not trust her terrors with my own;
Nor ev'n to Aldhelm yet avow our union.

I vii 1 *Bring in the Bishop*: an unfortunate line, the consequences of which are described by Thomas Campbell, drawing on the reviews of the play in London newspapers: 'At that time there was a liquor much in popular use, called Bishop . . . when jolly fellows met at a tavern, the first order to the waiter was, *to bring in the Bishop*. Unacquainted with the language of taverns, Miss Burney made her King exclaim, in an early scene, '*Bring in the Bishop!*' and the summons filled the audience with as much hilarity as if they had drank of the exhilarating liquor' (*Life of Mrs. Siddons*, 1834, p. 267).

I v 70 *So*: 'finely' is deleted.
I v 72 *which*: an insertion, replacing 'that'.
I vi 1 *Sir, –* : 'attends' is deleted.

Scene ix

EDWY.

Enter ALDHELM.

EDWY. Right welcome, Aldhelm. Now the Hour is come
Which to a point thy generous friendship draws;
Now, pious Aldhelm, by thy sage exertions
Complete the work of years: give to thy King,
Whose youth thou'st train'd, whose principles thou'st form'd,
Who owes to thee whate'er he dare approve
In the frail texture of his composition,
Give to him now that sweet domestic peace
Which fixes his felicity in virtue.
ALDHELM. All this, O King, if Aldhelm could bestow
His lot were blest beyond the lot of Man.
EDWY. Be blest, then, and make Elgiva my Queen!
Thy holy word will sanctify my choice,
And render it as popular as pure.
ALDHELM. The case is arduous, the success precarious;
Anxious I feel –
EDWY. Anxious? be rather proud!
Think of the worth 'tis giv'n thee to proclaim;
Think of the object who for justice calls;
Think of the purity 'tis thine to assert,
Unsullied as thy priestly Character,
And in fair harmony with all thy doctrines.
ALDHELM. Entranc'd, my Lord, in one absorbing passion,
You see not, hear not, heed not what surrounds it.
How happy while it lasts the fond illusion
Gave it not heavier weight to disappointment!
EDWY. Talk not of disappointment and illusion –
– Yet speak the worst at once, – I'll hear, and heed thee!
ALDHELM. Briefly and plainly then, prepare, O King,
For adverse times, nor yield your mind to hope
While all around is menacing disaster.
Dunstan, who forfeited by Edred's Death
His power at Court, preserves it with the people:
Replete with discontent, he waits the moment
To seize once more supreme authority. –

I ix 3 *pious*: an insertion, replacing 'noble'.
I ix 27 *at once*: an insertion, replacing 'you think'.
I ix 32 *His*: an insertion, replacing 'All'.

EDWY. He waits in vain. Am I not well appriz'd
What frauds he practic'd on my credulous Uncle?
Should he incense me, can I not expose him?
ALDHELM. That were a perilous and last resource.
Edred, all unsuspicious of imposture,
Confirm'd, with dying voice, his every act.
Archbishop Odo, weak and superstitious,
Who heads the Church, gives him implicit rule,
And thinks him Sainted, ev'n while yet on Earth.
EDWY. Alike unworthy each to raise alarm,
The one an Hypocrite, a Fool the other.
The Council call, and with the plain eloquence
Of simply stating truth, baffle them both.
Nor art, nor folly, stand that genuine test.
ALDHELM. Spoke as a King should speak! but yet, alas! –
EDWY. Banish, O Aldhelm, these depressing doubts,
Nor let cold fear thus chill thy liberal zeal.
ALDHELM. No, 'tis of warmer stuff than fear can chill!
My very prudence springs but from it's source.
My zeal, with fire perennial, dares all frost.
No stormy blast the changing times can blow,
Nor ev'n Man's Winter of decline, that hopes
No Spring's return, save Immortality's,
Can stagnate in these veins the genial current
Of faith, devoted loyalty, and love,
Which circulate in tides of youthful fervor,
Though rushing from a Heart Youth long has flown.
But here I see, my lord, the opposing phalanx
Of nearly the whole priesthood, led by Dunstan, –
EDWY. And what of that? what of their opposition?
Can they disprove the worth of Elgiva?
Can they conceal it from the nation's knowledge?
He who objects to own her for his Queen
Say to him simply this, in Edwy's name, –
Point out her Equal, – he'll resign his choice!
ALDHELM. Yet, sir, the canon-Law –
EDWY. 'Tis most unjust! –
ALDHELM. Odo and Dunstan –
EDWY. I disdain, despise them! –

I ix 46 *The Council call*: an insertion, replacing 'Open the Council'.
I ix 50 *depressing*: an insertion, replacing 'distrustful'.
I ix 54 *with fire ... frost*: an insertion, replacing 'defies all frost, with fire perennial'. Before 'dares', 'br' is deleted; 'braves' was probably intended.
I ix 57 *save Immortality's*: an insertion, replacing 'on this side Immortality'.
I ix 59 *devoted loyalty, and love*: an insertion, replacing 'respectful love, devoted loyalty'.
I ix 71 *ALDHELM*: 'The ‹insults› and power of' is deleted.

Suppose the worst, suppose them obdurate,
Can I not wed her, and defy them all?
ALDHELM. Great Heav'n forefend!
EDWY. And why? – what would ensue?
ALDHELM. Tumult! – Divorce! – Rebellion! –
EDWY. Ha! –
ALDHELM. My Lord,
Pardon my plainness while in Time I urge
The danger dire of such an enterprize.
EDWY. I fear it not! – but yet would be prepar'd.
If they resist I'll bear my suit to Rome.
ALDHELM. At Rome the Appeal could only be availing
If satisfied by the Church-Dignitaries.
EDWY. Not else?
ALDHELM. Not else.
EDWY (*disordered*). I fear I have been too hasty! –
ALDHELM. My Lord?
EDWY. – Thou think'st, then, it were difficult
Without the sanction of the Pope, t'oppose
The Council, if dissentient? – saids't thou so?
ALDHELM. Not difficult, my lord; impossible.
EDWY. Impossible? – my Head grows giddy! –
ALDHELM. Sir!
EDWY. No more, but call the council. Speed thee, Aldhelm,
Nor let one negative disgrace my honour.
Urge them to give me with my Crown my Queen;
Bring me their full, unanimous consent,
Bring me the Votes of my united people
That she who fairest stands for worth and virtue,
For pious principles, and royal Birth,
May justly share My Throne, and Their allegiance.
Exit.

Scene x

ALDHELM.

ALDHELM. How fair the trust of inexperience!
How pure the innate credulity of Youth!
Might they but last! – This Earth were then too blest. –
No – let them fly to Heaven! – be't our's to follow!

I ix 72 *the*: 'very' is deleted.
I ix 77 *danger dire*: an insertion, replacing 'dire dread consequence'.
I ix 83 *Edwy*: 'Nothing' is deleted.
I x 1–2 *How fair . . . Youth*: an insertion, replacing:
How beautiful is trusting inexperience!
How innocent, how lovely is credulity!

Scene xi

ALDHELM.

Enter ODO, DUNSTAN, REDWALD, CERDIC, ERNULF, *and* LORDS *of the Council.* ALDHELM *takes the chair as President, the rest sit round.*

ALDHELM. The King's express commands assemble us,
For special purposes. A weight of care
Hangs on his royal breast. His gracious pleasure
Requires from his good subjects, met in Council,
A fair attention to his difficulties,
And confirmation of their best solution.
Hard is the task assign'd to mortal man,
Severe though just, the forfeit of his sins.
To guide his erring, wander-loving steps
Firm through the rocky road of mazy life;
To force his Eye from the bewildering landscapes
Which to the right and left alternate court it;
Even this, the mere omission of offence,
Our Daily trial, and our common duty,
Exceeds the strength of fall'n humanity.
Yet harder, heavier still the Monarch's task.
As Man, born to endure the ills of Man,
As King, bred to assume himself above them.
Taught with elaborate glare to hide each feeling,
To deck out sorrow with composure's smile,
And with the semblance of complacency
To cover hoards of anguish: covering slight
That cannot mask the Heart, yet in the effect
Most ponderous, weighing down with force tyrannic
All Nature's fairest elasticities,
Which bent, not broken, with unceasing struggle
Strive to assert their genuine, pristine rights.
DUNSTAN. Efforts like these are plays of fortitude,
At which a manly mind disdains to murmur.
ALDHELM. Manly? No, Dunstan, rather say unfeeling;

I xi 1 *commands assemble*: an insertion, replacing 'command assembles'.
I xi 11 *force*: an insertion, replacing 'turn'.
I xi 12 *alternate*: an insertion, replacing an illegible word or words.
I xi 15 *the*: an insertion, replacing 'th'extremest'.
I xi 21 *of*: 'unmov'd' is deleted.
I xi 23 *the*: an insertion, replacing 'th''.
I xi 25 *Nature's*: 'finest' is deleted.
I xi 28 *Efforts like*: an insertion, replacing 'Exertions such as'.

Scornful, despotic, cynical the mind
Which prides itself on standing insulated.
Not such the Soul, though such the state of Edwy:
Mortal himself, he aims not to be lock'd
Against his fellows: but while he conscious shares
Their pains, afflictions, and infirmities,
He justly claims the equal privelege
To sooth Life's storms with chos'n alleviations.
That Edwy should avow his royal purpose
To grace his youthful Throne with a fair Partner,
Must give through-out his realm a general joy,
And meet each rational expectancy.
Strange therefore may it seem that thus to act
In consonance with policy and nature,
Should fill his mind with deep solicitude.

ODO. So young? already thinks the King of marriage?

ALDHELM. Not from his youth arises a demur:
Is he not Father of a mighty people?
And shall he shrink from a domestic charge?
No, 'tis distress of wider intricacy,
Affecting not alone his mental interest,
But, what he holds in awe, a papal Edict.

DUNSTAN. How? is there danger to the papal power?

ODO. This timely hint may put us on our guard;
Aldhelm, our thanks.

ALDHELM. Fear ye the noble Edwy?
He aims to soften, not defy the Law,
Your candour seeks, and deprecates your censure.

CERDIC. Who can fear Edwy? generous, great, and good,
His bounteous Heart is less his own than our's.

ERNULF. We grieve to hear our gracious King hath cares.

REDWALD. We're most amaz'd; so young, so full of power, –

DUNSTAN. Bishop of Winchester, if what ensues
May raise disturbance, or require reflection,
Bring it not forward yet, Time is too precious.
Attend we first to greater, higher rights,
Of pious call. The convent's reformation
Ill suits delay.

ALDHELM. For that we meet not now.
Wave to a future time all other interests,
And give the present to my high commission.

I xi 38 *alleviations*: two illegible lines are deleted.
I xi 55 *Aldhelm, our thanks*: an insertion, replacing 'We thank thee, Aldhelm'.
I xi 55 *ye the noble Edwy?*: an insertion, replacing 'not the noble Edwy:'.
I xi 66 *call*: an insertion, replacing 'import'.
I xi 68 *other interests*: an insertion, replacing 'Foreign business'. In place of 'business', 'questions' was first inserted, then deleted.

'Tis not alone a Queen our Monarch asks;
His liberal Heart demands congenial worth,
A Friend approv'd, a partner tried and known:
For through the wide domain a sovereign sways
Who like it's ruler needs that lenient balm?
His dignity o'erwhelms all confidence,
His pow'r tempts deceit, his rank, servility;
And sordid Envy ever seeks to drive
All bosom comfort from exalted stations.
Dark Misconstruction lies in secret ambush
To point each fault, and aggravate each doubt;
The Sovereign fears some mercenary aim,
The Subject dreads desertion, or caprice,
Till, dup'd alike, Patron and Patronis'd
End their suspicious commerce in disgust.
A subject, thus beguil'd, flies to his Equal; –
But, if betray'd, where may a Monarch turn
And hope a happier choice?
CERDIC. Who can reply?
ALDHELM. If one there be to whom his Heart may open,
Of similar interests, simpathetic feelings,
Whom he may safely cherish, blameless trust,
Uncheck'd by fancied danger's dark suspicion,
Which can discolour justice, truth, and honour
Must it not be – his Wife? –
DUNSTAN. His Wife? –
ODO. Amazement!
ALDHELM. Where lies the wonder? if his choice be worthy,
If, sprung from royal blood, her Birth be equal,
If, bred with pious care, her mind's endu'd
With each right principle that from the Throne
Should spread it's radiant influence o'er the nation,
And ope each latent source of dormant virtue; –
If, – if, in brief, – a Princess such as Elgiva –
DUNSTAN (*rising*). What do I hear?
ODO. How? Elgiva? –
DUNSTAN. His kinswoman? –
ERNULF. It cannot be.
REDWALD. The Law allows it not.
ALDHELM. Attend! He strives not to infringe the Law,
He knows himself it's guardian: all his purpose

I xi 70 *asks*: 'To fill the English Throne were little difficult' is deleted.
I xi 75 *dignity o'erwhelms*: an insertion, replacing 'elevation overwhelms'.
I xi 76 *tempts*: an insertion, replacing 'invites'.
I xi 83 *Patron and*: an insertion, replacing 'the Patron and the'.
I xi 87 *Who can reply*: an insertion, replacing ''Tis truly difficult'.
I xi 92 *justice, truth, and honour*: an insertion, replacing 'ev'n the fairest seeming'.

Is simply to appeal –
DUNSTAN. Break up the Council! (*all rise*.)
Let us not sit to hear a proposition
Bordering on blasphemy 'gainst the Roman see.
ALDHELM. Let me explain. Th'approaching Coronation
Urges the King –
ODO. 'Tis dangerous to listen! –
DUNSTAN. Break up! – Most pious Odo, lead the way! –
ODO. First let the coronation be perform'd
With loyal unanimity; that over,
Summon a Council Ecclesiastical,
And weigh the matter with becoming strictness.
DUNSTAN. Be it so.
LORDS. We're all content.
DUNSTAN. And now adjourn
Till our young King be crown'd.
ALL. Adjourn, adjourn!

Exeunt in disorder.

Scene xii

ALDHELM.

ALDHELM. How stubborn is dark prejudice! – O Edwy!
How will thy spirit brook what it must bear,
Keen disappointment, and repulse indignant?

I xi 107 ***Roman see***: the authority of the Pope; the Papacy.

I xi 107 ***see***: an insertion, replacing 'Pontiff'.
I xii 1 ***How stubborn*** ... ***Edwy***: an insertion, replacing an illegible line.

ACT II

Scene i

An Apartment of ELGIVA'S.

Enter ELGIVA *and* ELTRUDA.

ELTRUDA. Must then the sacred Honours of this Day
Which give the King his crown, the Land it's Ruler,
Bless each inhabitant of England's realm
Except the first and best, it's lovely Queen?
ELGIVA. Risk not, Eltruda, that alarming sound;
No Queen is Elgiva, – O were that all!
ELTRUDA. Canst thou thus mildly see thyself bereft
Of each appendage to pre-eminence?
Canst thou, while thus from prejudice denied
Thy conscious rights, relinquish them unmurmuring?
ELGIVA. Eltrud, those conscious rights alone support me;
They shield me from the piercing shafts of Censure,
Teach me to bear disgrace unmortified;
Brace every trembling nerve, inspire my courage,
Re-kindle dearest hopes, and, – more than all! –
They keep me innocent, – yet give me Edwy!
ELTRUDA. How sweet is patience struggling with affliction.
Yet why, my Princess, seek this lone Retirement?
Haste to thy friends, and turn thy lingering Eyes
From objects that awake but disappointment.
ELGIVA. No visionary hope attracts me hither;
Full well I know 'twere vain, this splendid Day,
To breathe an expectation of my lord:
Devoted be it to his kingdom's pomp.
Yet wonder not, Eltruda, if I haunt
With all-resistless impulse, the lov'd spot
Where grateful simpathy first touch'd my Heart.
Here Memory preserves my happiest records,

II i 7 *mildly*: an insertion, replacing 'gently'.
II i 8 *to*: an insertion, replacing 'fair to just'.
II i 12 *piercing*: an insertion, replacing 'pitiless'.
II i 13 *Teach me to bear*: an insertion, replacing 'And teach me to endure'.
II i 14 *Brace every trembling nerve*: an insertion, replacing 'They brace my trembling nerves'.
II i 15 *Re-kindle dearest hopes*: an insertion, replacing 'Rekindle every hope'.

For Edwy here nam'd me his honour'd choice;
Here Retrospection waves a magic Wand
That conjures back sweet Hope's elastic charm;
Here native spirits hold their pristine sway,
And respiration seems inhaling joy!
ELTRUDA. Such happiness, befitting thy fair mind,
Suits well the upright tenor of thy life.
Yet, Princess, –
ELGIVA. Heavens! I hear my Edwy's step! –

Scene ii

Enter EDWY, *crowned, and in his Coronation Robes.*

ELGIVA. Is't possible, my Lord? – is't not illusion? –
Am I so happy? – is't indeed no dream? –
How hast thou stol'n from state this one kind moment? –
How shall I thank thee that thou makst it mine?
EDWY. Thou fair assemblage of all female excellence!
Thus canst thou greet me, with this generous warmth,
Those open smiles, those Eyes of radiant kindness,
When thou beholdst me powerless to sustain
Thy rights, and issuing fresh from unshar'd honours?
ELGIVA. Ere thou canst make a merit, my lov'd Lord,
That honours such as those I lightly lose,
Thy Heart from Elgiva must be withdrawn;
And then, bereft of that, – if then she lives! –
She may find time to prize thy meaner greatness.
EDWY. O what a welcome this to still the Storm
Of rough'ning passions beating in my breast,
Where each indignant feeling to offend
Or king or Man, rages in wild commotion.
Dunstan, assuming more than regal power,
With unexampled insolence, broke up
The Council, ere My message was receiv'd,
Or Aldhelm could collect the general voice.
I had ne'er submitted to this lone solemnity, –
– Lone without Thee, midst crowded multitudes,
Distasteful, wearisome! – had I not fear'd
Refusal might have waken'd some suspicion
Touching our marriage.
ELGIVA. Ah! great Heaven forbid!

II ii 1 s.d. *his*: an insertion.
II ii 3 *kind*: an insertion, replacing 'short'.

If at the mere suggestion of th'intent
They started thus repugnant, – O my lord,
What have I not to dread when all is known?
EDWY. Dread nothing, my lov'd Elgiva; our secret,
Entrusted to so few, can ne'er transpire:
Nor will I publish it, till All unite
To own thy worth, and ratify its claims.
ELGIVA. But was it safe, my lord, to quit thy Nobles?
Will they not wonder – ? Must thou not return?
EDWY. I shall be with them ere a doubt is rais'd.
They're unacquainted with the secret avenue
That brings me hither. Leofric alone
Attended me; and, faithful, guards the entrance.
Even Aldhelm is a stranger to these meetings;
He knows not thou art mine, and would condemn them.
O let me then partake a few short instants
The happiness of Innocence and Peace.
Those I have left are Strangers to such bliss,
Nay doubt, perchance, it's possible existence.
Noisy festivity, repasts luxurious,
Spirits inflam'd by Reason's conflagration,
Loud, frantic, vehement, discordant sallies
Of ev'ry passion of th'uncultur'd mind,
Frame their fierce joy, and mark their furious tempers.
O what a soft refreshment to escape
From such disorder'd scenes of sensual riot
To purity and elegance like thine!
How do they distance vain magnificence,
Whose baseless pleasures charm but vacant minds? –
ELGIVA. Should I not blush to avow, thus deep-indebted,
I yet can hope to owe Thee, Edwy, more?
Hope thou wilt still preserve this partial goodness
When all that first excited it is flown:
When Youth deserts, when Spirits wing their flight,
When Health, and all Life's powers are on the wane,
And Gratitude alone remains unfaded!
EDWY. Fear not, sweet Elgiv, thou canst lose thy hold;
Have I not loved Thee from my earliest dawn?
My Reason and my passion, like fair twins
Grew in thy sight, and fasten'd on thy charms.
All that my opening faculties acquir'd
Nourish'd my fervent fondness with approvance;
All I was taught of virtue, self-controul,
Delight in doing good, – constant I saw
Exemplified in Thee. Thou wert my Lesson;

II ii 66 *like fair twins*: an insertion, replacing 'hand in hand'.

The test by which I tried Instruction's justice,
My proof for truth, my energy for action!
Thus in th'extremest youth thy beaming virtues
To Me were purest elemental light.
O loveliest Pattern! sweet unconscious Monitress!
How often has my Heart – what do I see? –

Scene iii

EDWY. ELGIVA. ELTRUDA.

The principal Door suddenly opens, and DUNSTAN *rushes in, followed by* ODO, REDWALD, ERNULF, *and* NOBLES.

DUNSTAN. Nobles advance! Archbishop! Redwald! Ernulf!
Enter! and view! – Were my suspicions false?
See! see! behold my prophesy confirm'd,
See England's King in base seduction's arms!
EDWY. Great Heaven! –
ODO. Most righteous Dunstan, 'tis too plain!
REDWALD. This I had ne'er believ'd!
ERNULF. Surprise confounds me!
DUNSTAN. Blush, blush O King! a blot like this will fix
On the first records of thy opening reign
A stain indelible.
EDWY. Am I awake?
I know not if I see, I hear, I live! –
DUNSTAN. The mists of lawless passion cloud thy senses.
EDWY. No, daring Dunstan, 'tis thy boldness stuns them!
Presumptuous man! presumptuous Nobles all!
To dare intrude on a retreat like this
Sacred to the fair fame of female honour.
DUNSTAN. Well mayst thou covet secresy and darkness;
The World, with all its crimes, and foul corruptions,
Would yet start back with horror from this Sight.
EDWY. Dunstan, by Heaven –
ELGIVA. Oh take me hence, my lord!
Save, shield, protect me from a Scene like this!
EDWY. Sweet Elgiv! chaste as lovely, – pure as fair! –

II ii 75 *beaming virtues*: an insertion, replacing 'rising excellence'.
II ii 76 *were*: an insertion, replacing 'was'.
II iii 2 *and view*: an insertion, replacing 'behold'.
II iii 3 *See! see ... confirm'd*: an insertion, replacing 'Come, see the confirmation of my prophesy'.
II iii 8 *records*: an insertion, replacing 'annals'.

Ill, ill indeed a scene like this accords
With the soft texture of thy timid nature.
Yes, I will take thee hence, nor for a moment
Degrade the dignity of innocence
By suffering calumny to breathe upon it.

leading her away.

DUNSTAN. How! in the very face of all thy Court,
And ev'n in presence of the mitred Odo
Dost thou indulge this licence to dishonour?
This privilege to dissolute pursuits?
EDWY (*returning*). O more than patience, more than man can bear!
Dunstan, thou cleav'st the secret from my Soul
Which sooner –
ELGIVA. Oh my lord – first let me go! –
EDWY. I will! I am to blame. –

again taking her away.

DUNSTAN. Together go ye?
Canst thou not bear to send her from thy sight,
Nor spare one solemn Day from guilty love?
EDWY (*quitting* ELGIVA). Ha! guilty! – hear then, Nobles –
ELGIVA. O forbear!
Forbear, my lord! – let me retire alone, –
And name me not. –
ODO. That were indeed most fitting.
EDWY. Not name Thee? Yes, while voice is lent to Edwy
Thy name shall vibrate through the passing Winds,
And with thy virtues fill the circling air! –
DUNSTAN. Nobles, observe, all crimson'd o'er with shame
His very concubine –
EDWY. How? Concubine? –
May I be blasted with perpetual infamy
If I bear this! if I defer a moment
To blazon to the World – She is My Wife!
ALL. Ha! – Wife? –
ELGIVA. Ah me!
ELTRUDA. My Princess! – Help! She faints! –

II iii 44 *concubine*: Hume writes that the monks readily used 'the name of *concubine*, or other more opprobrious appellation' for the wives of their rivals, and that Dunstan 'probably bestowed on the queen the most opprobrious epithet that can be applied to her sex' (*History*, I, 93, 95).

II iii 42 *air*: four illegible lines are deleted, one concluding Edwy's speech, one spoken by Elgiva, and two by Edwy.
II iii 44 *EDWY*: 'Concubine' is deleted.
II iii 48 *My*: an insertion, replacing 'O'.

EDWY. What have I done? – Look up, my Elgiva!
My Wife! My Queen! – Oh rash precipitance, –
ELTRUDA. See, she revives –
EDWY. My Elgiva! –
ELGIVA. Ah Heaven! –
EDWY. Come, let me take Thee from this daring tribe. –
Eltruda, help, – My Lords, expect me back! –
Forgive me, Gentlest, this too rough assault
For nerves and sensibility like thine.
On Me, My Elgiv, lean. –
Exit ELGIVA, *between* EDWY *and* ELTRUDA.

Scene iv.

ODO. DUNSTAN. REDWALD. ERNULF. NOBLES.

ODO. This passes credence!
REDWALD. Married? – without consent of Church or State?
ERNULF. When, and by whom has this bold act been done?
DUNSTAN. Assuredly 'tis Aldhelm's. Hear me, Lords!
Quick let it be repeal'd; condemn the Offenders
To separation and severest penance
Lest signal judgement on our heads should fall!
ODO. Most just.
ERNULF. We're all agreed.
REDWALD. Yes, part them! –
ALL. Part them!
DUNSTAN. If Edwy should resist – which Grace forbid!
For I would fain not think him culpable
Of guilt deliberate; – should he dare resist,
Far other means than menace, or persuasion,
Must crush this most flagitious intercourse.
ODO. Dunstan, thy dictates shall direct us all.
DUNSTAN. 'Tis meet that some few friends to papal rights
Assert their awful cause. Should we submit
To see them ravaged of authority
What might we not expect in issue? chief,
The Convent's great Reform would be disputed.
Unite we to impede that sacrilege.
ERNULF. I well remark'd when Aldhelm was address'd
To aid us in this fair surprise, he seem'd
Far more incens'd to put the King to the proof

II iii 50 *Oh*: an insertion, replacing 'pardon my'.
II iv 17 *ravaged*: an insertion, replacing 'thus divested'.
II iv 20 *we*: 'firmly' is deleted.

Than to believe him criminal.
REDWALD. Cerdic, too,
Refus'd to join our party. When I urg'd him
Forthwith to follow, and expose the King,
If here, as we suspected, he retir'd
Licentiously to wile away his time;
When this I press'd, he proudly turn'd upon me,
And cried: Do what you list, my lord of Devon;
For me, I ever scorn'd to force myself
With prying Eye intrusive, on my Equal;
Shall I then do it on my lord the King?
DUNSTAN. Brave Redwald, 'twas most insolent, and calls
For merited resentment.
REDWALD. And shall have it!
DUNSTAN. Cerdic I've mark'd; Friend of the Seculars,
And Aldhelm's Friend.
ERNULF. The King returns.

Scene v

Enter EDWY.

EDWY. Yes, Nobles,
Your King returns; collected, though amaz'd,
And ready now to meet each bold assailant.
No deed like this for savage violence
Has stain'd our records, or disgrac'd tradition,
Since Man was leagu'd to Man by social compact.
That Me, your Sovereign, you thus dare pursue
Ev'n on the Day when every legal tie
Invests Me with acknowledg'd power superior,
I yet might pardon to the haughty temper
Of Jealous Chiefs, unbending to controul:
But thus to violate the chaste abode
Of female purity; to burst the bond
Than Law more sacred, and than Force more potent,

II iv 33 *on*: an insertion, replacing 'to'.
II iv 36–37 *DUNSTAN ... Friend*: an insertion, replacing:
Let him look to't!
DUNSTAN. 'Tis well resolv'd. This Cerdic
Is the fast friend of Aldhelm and the Seculars;
I've marked him long.
II v 13 *Of female purity*: an insertion, replacing:
In which, of Blood as pure as royal, dwells
The highest of her Sex.

Of Nature's first, instinctive sense of honour
That fosters and protects whate'er's defenceless;
This is a breach of moral rectitude,
And each prescriptive decency of Life:
And deeply though I feel it as a man,
The general good calls for severer wrath
From your first Magistrate's insulted dignity.
ODO. We seek, Sir, your true glory, and right Welfare.
ERNULF. Led on by pious Dunstan, Sir, we follow'd
T'investigate the cause why our young Monarch
Forsook us, ere we well had vow'd allegiance.
REDWALD. We meant not to offend, but came offended.
Was't not injurious, Sir, to quit the Banquet,
Mocking thy Nobles with an Headless Feast?
DUNSTAN. Aptly that sovereign pleads his dignity,
And raves of justice, decency, and order,
Whose conduct fits his sentiments; who ne'er
By passion sway'd, by pleasure call'd aside,
Revels in Love licentious. –
EDWY. Dunstan hold!
By Heavens I will not hear another word
That breathes such blasphemy! Licentious Love?
Have I not loudly own'd her for my Wife?
Did you not hear? Would you that I repeat it?
Hear, then, ye Nobles! hear Me, mitred Odo!
And thou, proud Dunstan, hear! She is My Wife!
Hear it, O Earth! and all ye Sons of Men!
Hear it, high Heaven! – Yes, there it stands recorded
In the pure vows of her seraphic faith! –
She is My Wife! – The Princess Elgiva! –
Now wilt thou tell me of licentious love? –
My married, legal, and most virtuous Wife! –
And whom I make My Wife – becomes Your Queen!
DUNSTAN. Rejoice, O King, thou speak'st the thing that is not:
It were not valid. Thou art safe. She's – lost.
EDWY. She lost? – Unsay those Words! – Wretch! on thy life
Touch not Her Fame!
DUNSTAN. 'Tis thy pursuit that blasts it. –
EDWY. I bear not this! Calumniating Monk!
Clear Her, – or know –
DUNSTAN. Remove her from thy Sight;

II v 17 *a breach of*: an insertion, replacing 'an injury to'. In place of 'This is an injury to', 'Is an offence to' was first inserted, then deleted.
II v 22 *We seek ... Welfare*: an insertion, replacing 'We seek t'exalt, Sir, not insult thy dignity'.
II v 29 *Aptly*: an insertion, replacing 'Right aptly may'. The verbs 'pleads' and 'raves' are accordingly changed from 'plead' and 'rave'.

'Twere impious, henceforth, but to look at her.
EDWY. Impious to look at her?
DUNSTAN. To both perdition.
EDWY. I'll hear no more! Avaunt! – Nobles, retire! –
But list not to that fiend! – Dunstan! yet stay! –
If thou wilt cleanse the offended atmosphere
Of this most foul aspersion –
DUNSTAN. No, my lord;
She is undone. Take heed for her Undoer!
EDWY. Haste from My Sight! – My Lords – dispense with forms!

Exeunt all but EDWY.

Scene vi

EDWY *alone*.

EDWY. What fearful phantoms crowd upon my Brain!
Have I done wrong? – Misled My Elgiva? –
Mine? And how mine? – Were the rites competent
That bound us? – Heaven and Earth! if I've incurred
The Shadow of a blot on that fair name,
Though her pure Will and Nature 'scape the infection
How shall I ere recover peace for her,
Or for myself forgiveness? Terrible thought!
Have I betray'd her sweet docility?
Deceiv'd her Reason? and beguil'd her Love? –
Repentant horrour creeps through every vein: –
– O Elgiv! I will fly thy dangerous Sight,
Nor listen to thy voice, nor speak to thee
Till I obtain the sanction of a Synod
To ratify our Union. –

going.

II vi 3 *Mine? and how mine*: an insertion, replacing 'My Elgiva? – How? Mine?'.
II vi 4 *if I've incurred*: an insertion, replacing:
have I a doubt
Yet live? – O if there be incurr'd
II vi 10 *Love*: two lines are deleted:
Fatal precipitance! I should have waited
The Pope's award, and Aldhelm's sage concurrence.
II vi 11 *vein*: three lines are deleted:
To risk th'eternal exile of serenity
Thence, from its softest residence, her bosom,
That seem'd it's nat'ral Element. – How attone? –
II vi 12 *Sight*: an illegible line is deleted.
II vi 15 *Ah*: an insertion, replacing 'Great'.

ELGIVA (*within*). My Lord! –
EDWY. Ah Heaven! –

Scene vii

EDWY. *Enter* ELGIVA.

ELGIVA (*entering*). Still art thou here? My Kind and faithful Edwy!
I haste to set at rest thy feeling Heart:
I am better – I am well. –
EDWY. I am most rejoiced –
Adieu! –
ELGIVA. My Lord!
EDWY. I am summon'd hence – Farewell! –
ELGIVA. Go, then; I come not to detain but speed thee.
I know – but werefore thus avert thy Eyes?
EDWY. Heed me not, Elgiva, – I am not well –

going.

ELGIVA. Ill, and not heed thee? – Nay, a moment stay! –
How art thou ill, my Lord, in Mind, or Health?
EDWY. I prithee ask not, Elgiv –
ELGIVA. I submit! –
Ah! can obedience be a task with Thee?
I will not beg a word – a glance – a thought –
Only be well! –
EDWY (*turning suddenly towards her*). Impious to look at her! –
O fair – dread Object of my condemnation!
How look at ought beside! – Ah! fly Me, Elgiv! –
ELGIVA. Fly Thee, my Lord?
EDWY. 'Twould be most generous;
Fly, – for I cannot! –
ELGIVA. Nay then, let me go, –
Yet whither fly? – what path, what course pursue
To calm thy mind! – O Lord of all my wishes,
Look not thus strangely on thy wedded Wife! –
Am I not thine? – why dost thou start? – what means
That dark, black horrour gleaming from thy Eye-balls? –
Are those the Eyes – O can they be the same
So lately sparkling with delighted Love,
Gladd'ning my Heart with more than infant glee,

II vii 10 *Elgiv*: five illegible lines are deleted, two spoken by Elgiva, one by Edwy, one by Elgiva, and one by Edwy.

And bright'ning all around with grateful joy?
Can they –
EDWY. I pray no more! –
ELGIVA. O pardon me!
I had forgotten what I just had promis'd.
In silence I'll depart.

going.

EDWY. No, – better stay; –
Let Me, as 'tis most fit, depart myself,
And quit the fascination of thy dwelling
Ere terror and repentance – O adieu! –
Bear with a short suspence – support thy courage, –
Be all thou yet hast been, patient and noble! –

Exit.

Scene viii

ELGIVA *alone.*

ELGIVA. Gone! – Can it be? – without one balmy look
Of fostering tenderness to sooth his absence? –
Gone? – Has he left me thus? – all unappriz'd
Of what he purposes? – all lost in mystery? –
Abandon'd to the wildness of conjecture,
Fearing to hope, though starting from despair!
How fondly had I thought my Edwy's Soul
The faithful, safe deposit of this bosom,
Where all in sacred silence dwelt his cares;
I thought myself the partner of his mind,
Not to relax his strength, but share his pains,
Not to cast off, but mitigate his ills: –
Dear flattering error, whither art thou flown?
Did he not speak of terror and repentance? –
O Edwy! – is it, can it be from Thee,
Noblest and best! – I deprecate unkindness?

Scene ix

ELGIVA. ELTRUDA.

ELTRUDA. My Princess! all in haste the King has sent me
To watch and guard Thee: Wherefore? what's the alarm?

II ix 2 *alarm*: an illegible line is deleted.

Fast rushing onward as he spoke, he bad me
Seek, and support Thee.
ELGIVA. O reviving hearing!
There was he all himself! tender, humane,
Benignant! There his genial Temper beam'd
Bright through the thickening mists of hazy care
That dim'd the radiance of his mind's transparency!
ELTRUDA. He look'd, indeed, distress'd past utterance.
ELGIVA. O were he not distress'd, yet thus could leave me
How might I bear such unmixt wretchedness?
What but his constancy can now sustain me?
ELTRUDA. Doubt not of that; his love is unexampled.
ELGIVA. And yet – he spoke of terror and repentance –
Could I suppose that but to call me his
Were adequate to dissolve the Heart-dear bond
Of previous union, which first made us one,
Vain were all else to me! That bond dissolv'd,
What after-rite, what solemn Ceremony,
What awful invocation at the Altar
Could breathe sweet harmony o'er repented vows,
Or satisfy the Heart? The sacred knot
Which elevates Affection into Duty,
And sanctifies the fondest tenderness –
O such I hop'd had plighted me with Edwy –.
ELTRUDA. And canst thou fear thy fulness of success?
ELGIVA. I think not; – yet – why should he name repentance?
What anguish must have rent his generous soul
Ere he could speak that word! A soul so liberal,
So tun'd to vibrate to Affliction's touch,
So adverse to offend – O what of pain
Must Edwy suffer, ere he could inflict it!
ELTRUDA. Perhaps some insult of the haughty Dunstan –
ELGIVA. I know not what has pass'd, and dread to hear;
He could not speak; and now I grieve I urg'd him.
I should have stay'd his Time, and felt secure
Mildness for me alone constrain'd a silence
To him, perhaps, most sad. He ever spurn'd
All thought of self for others. Can I forget
How late he trembled at his dawn of power
With virtuous fear, lest it should be misus'd
Through his deficiencies? Is not his Honour –

II ix 16 *Were*: 'name' is deleted.
II ix 24 *the*: 'softest' is deleted.
II ix 25 *O such ... Edwy*: an insertion, replacing 'That was the knot I thought to tie with Edwy'.
II ix 29 *liberal*: an insertion, replacing 'delicate'.

ELTRUDA. Methinks I hear some sound! –
ELGIVA. O whence? – which way? –
ELTRUDA. There, up the secret passage.
ELGIVA. Then 'tis Edwy! –
O my lov'd Edwy! – (*runs to the private Door.*) No! – he is not here! –
I thought perchance he had hasten'd to relieve
My mind's affright.
ELTRUDA. Again some step approaches! –
ELGIVA. Nay, then it must be his! O joy of Heart!
Edwy! my Edwy! – why so slow? – Hist! listen!
Dost thou not hear a buzz like whispering? –
Perhaps 'tis some command for Leofric.
None other know this path. – My dearest Lord! –
Art thou not coming? – Ah! – Great Heaven! – Dunstan! –

Scene x

Enter DUNSTAN, *with three* RUFFIANS.

DUNSTAN. Shriek not – make no resistance – and you're safe. –
ELGIVA. Merciful Powers! –
ELTRUDA. What does this mean? –
DUNSTAN. Be still! –
The least disturbance robs you of your lives.
'Tis from perdition I would rescue you.
Prevent their cries, and seize them both at once. –
ELGIVA. O Edwy! – O my lord! – protect –
ELTRUDA. O hear me!
ELGIVA. O Dunstan by whate'er thy Soul has priz'd –
See at thy feet – alas! no haughty Queen –
A Suppliant Wife but sues her Lord protection –
An innocent female –
DUNSTAN. Off with her unheard
ELGIVA (*struggling*). Spoil me of Splendour, titles, rank and power –
Spare but my fame; blight not my untainted honour,
Wrest, wrest me not from Him whom every tie
Divine and human –

II ix 46 *perchance*: an insertion, replacing 'perhaps'.
II ix 49–50 *Hist! listen! ... whispering*: these lines were originally given to Eltruda.
II ix 51 *Perhaps 'tis*: an insertion, replacing 'That may be'.
II x 5 *once*: '*The Ruffians seize them*' is deleted.
II x 7 *priz'd*: '*The Ruffians gag, and carry them off*' is deleted. The remainder of the scene is an insertion.
II x 9 *her*: an insertion, replacing 'a'.

DUNSTAN. Wherefore list ye? Gag her;
She's but a Courtezan.
ELTRUDA. Foul, barbarous, slander!
ELGIVA (*still struggling*). Where art thou Edwy? – Hold ye gentle Soldiers!
Kind, pitying, good! – Forbear –
DUNSTAN. Away, away,
Pernicious Concubine.
ELGIVA. Where art thou Edwy?
Good people hear! I am your monarch's Wife!
Rend me not from his arms – Where art thou, Edwy? –
My legal lord! – my wedded Husband! – where –
The RUFFIANS *gag and force them off;*
DUNSTAN *silently following.*

ACT III

The Council Chamber.

Scene i

ODO. DUNSTAN.

DUNSTAN. Archbishop, she is safe. Repress all fear.
Who'll dare accuse me? Take this interval
Before the King and council meet, to arrange
What next is to be done.
ODO. Form thy own plans,
Most righteous Dunstan: what are thine, are mine.
DUNSTAN. With vigour, then dispatch we our Reform.
If Edwy prove intractable, – his Brother –
The young and gallant Edgar – is our Friend: –
Would he were more! –

II x 14–15 *Gag her* ... *Courtezan*: the reviewer for *The True Briton* cited this as an example of language 'inelegantly familiar, or ridiculously absurd' (23 March 1795).

II x 21 s.d *following*: a two-line scene is deleted:

Scene xi

DUNSTAN *alone*.

DUNSTAN. 'Tis done! in Ireland they may both be safe
Till Times are ripe for more security.

In Hume's *History*, Elgiva is taken to Ireland before she returns to her death in England.
III i 2 *Take*: an insertion, replacing 'Let us seize'.

ODO. Sayst thou?
DUNSTAN. If impious Edwy –
Deaf to our exhortations, brave his fate, –
If he resent losing his concubine – –
ODO. Speak out! why hesitate?
DUNSTAN. If he burst forth
With rage ungovern'd –
ODO. Ha! –
DUNSTAN. Which Heav'n forefend! –
But yet –
ODO. Proceed! –
DUNSTAN. I grieve – but if't be so –
Or Dunstan perishes – or Edgar reigns!
ODO. Then Edgar reigns! – for Dunstan must be sacred.
DUNSTAN. With pain I point out the alternative:
But what remains? Edwy will bear no yoke,
Not ev'n Ecclesiastic. Headstrong and bold,
He rules himself, or listens but to Aldhelm,
Who would forego each priestly privelege
To cast the Land on civil Government.
Edgar is better fram'd to serve our cause.
Contented with his own pursuits, the burthen
Of State affairs on us would soon devolve.
ODO. 'Tis true; But might the people not resist?
Would the crowd comprehend our arguments?
DUNSTAN. The crowd require illusion, not conviction.
Entic'd by Terror, caught by Ambiguity,
Weakly beguil'd, and eagerly amaz'd,
To them 'twere useless to discuss opinions;
They must be led more vigourously to action
By calling forth their passions, and their interests,
By raising fears unnam'd, and Hopes mysterious.
This task be mine – if Edwy's doom'd to fall:
But yet I hope –
ODO. No; fall he must; 'twere better
Than let Impiety prevail.
DUNSTAN. I yield, – –
Though most reluctant. 'Twere an easy matter
To lure him to his ruin. Well I know

III i 11 *losing*: an insertion, replacing 'the bearing off'.
III i 16 *sacred*: an illegible line is deleted.
III i 17 *With pain*: an insertion, replacing 'Unwillingly'.
III i 19 *bold*: an insertion, replacing 'hardy'.
III i 21 *Who*: an insertion, replacing 'And Aldhelm'.
III i 24–25 *the burthen ... on us*: an insertion, replacing:
on us
The burthen of the State

His sanguine, rash, and irritable spirit,
And to the quick could stir it with a word.
But first I'll try the soothing of persuasion;
Then menace fierce; and, lastly, malediction:
If, after that, he spurn our sage Reform –
But hold! – Aldhelm – –

Scene ii

ODO. DUNSTAN. ALDHELM.

ALDHELM. Has yet the King appear'd?
DUNSTAN. 'Twould better suit with us, my good Lord Bishop,
To aver this harmlessness of ignorance
Than with the known director of the state,
The favour'd Aldhelm. Who so well can say
Where Edwy is, or what are his intents?
ALDHELM. The noble Edwy both deserves and holds
The Heart of Aldhelm. True, he has favour'd me;
But I have sought not to disgrace his honour
By fouling the fair springs of Gratitude
With Adulation's mire.

Scene iii

ODO. DUNSTAN. ALDHELM. SIGEBERT.

ALDHELM. Sigebert, thy news?
SIGEBERT. Soon as th'Attendants miss'd the Princess Elgiva
They hasten'd to reveal the dreadful tydings.
The King was found immers'd in anxious care
Close with the Priest who tied the fatal knot,
Consulting on the means to call a synod
For rendering it more permanent.
DUNSTAN. O horrour!
ALDHELM. How did he bear the shock?
SIGEBERT. Most grievously.
And all unarm'd, unaided, unattended
Forth from the Gates he rush'd. His fruitless search

III ii 3 *To aver ... ignorance*: an insertion, replacing 'To make enquiry, and aver incertitude'.
III iii 4 *anxious care*: an insertion, replacing 'weighty business'.

Quick drove him back enrag'd. And now in person
He issues every order for pursuit.
Meanwhile, 'tis his command th'assembled Council
Should here await his pleasure.

DUNSTAN. Odo, mark!
Th'assembled Council may await his pleasure
While his voluptuous Courtezan detains him!

ALDHELM. Is't Elgiva thou styl'st voluptuous Courtezan
Dunstan, this calumny as ill befits
The mild forbearance of thy holy order,
As the fair Character it thus traduces.
The Princess –

DUNSTAN. What! does the immaculate Aldhelm
Grant his protection and defence to Elgiva?

ALDHELM. Her Youth and inexperience claim protection;
Her innocence, defence.

DUNSTAN. Thou seemst with pride
To boast thyself her Champion?

ALDHELM. Werefore not?
The grateful task of justifying virtue,
The pure delight of raising the oppress'd,
And mitigating evils undeserv'd, –
This is a pride we need not blush to own,
An energy that well becomes our calling,
And casts a lustre o'er the general darkness
Of grave austerity's habitual gloom.

DUNSTAN. And flow such sentiments from Edwy's Tutor,
Who should have early taught his Prince to know
That Royalty exists but for the State?

ALDHELM. How hapless then its chief, if there's no period
He may unbend to Nature, and to Truth?
If He, alone, must ever be denied
From Grief all solace, all release from art,
From Home, Man's dearest right, all trust and comfort!

Scene iv

ODO. DUNSTAN. ALDHELM. SIGEBERT.

Enter CERDIC. ERNULF. REDWALD. LORDS.

ALDHELM. My Lord of Mercia, com'st thou from the King?

III iii 35 *That Royalty ... State*: an insertion, replacing 'That thoughts of State alone accord with royalty'.

CERDIC. Yes; and I have left him in a state to envy
His lowliest, meanest subject; while on him
The wretchedest might cast an Eye of pity!
Nothing he learns, while every where he flies.
ALDHELM. Tell me the way that I may follow him? –
CERDIC. All are dismiss'd; he will not be attended.
Barbarous his sufferings! Most accurs'd the Wretch
Who has inflicted them! –
DUNSTAN. Sayst thou accurst? –
Dost thou suspect –
CERDIC. No one. A generous nature
Scorns to give Cause, or entrance to suspicion.
My malediction fastens on the fact,
And, – be he whom he may, it's perpetrator.
ODO. Unknowing is thy speech, and all too hasty.
The deed devout of parting lawless Lovers
May save the destin'd Land from misery.
DUNSTAN. If there be any curse, who merits it
But he, the abettor of this impious union,
It's Patron, it's encourager? –
ALDHELM. If Aldhelm
Seem in thy mind to answer this description,
Know, Dunstan, thou art misled. Pitying his youth,
So early cast on the sharp Rocks of royalty,
Whose perils distance all his help they need,
I wish'd his cares, indeed, a virtuous Partner:
But, on my faith, I knew not of the marriage.
ODO. What ills are his who may command a Kingdom?
ALDHELM. His Rank superior tells thee what he lacks,
Equality; which Wedlock solely gives.
There he might taste fair Friendship's purest essence, –
Friendship, next Virtue, Heaven's best gift to Earth.
DUNSTAN. Whom should he seek, on whom bestow his friendship
If not on those with holy rites invested?
Here, in the priesthood, let him find his solace.
ALDHELM. I grant the general doom, that leads one way,
And Man's brief exit, might instruct us all
To seek our common refuge in Religion:
But life, so various in it's course, requires
Various resources 'gainst tempestuous Days.
The Monk may live to solitude and prayer;
But he whom duty calls to public office
Will find such countless, deep solicitudes,
Rolling, successive, in an endless tide

III iv 41 *countless*: 'wasting' is deleted.
III iv 42 *an*: an insertion.

Which pours, through crossing Channels, to his Heart,
That to allow no vent might overwhelm it.
DUNSTAN. Wouldst thou have Edwy trust a female Breast
With state transactions? Wouldst thou –
ALDHELM. There were no need;
Rest that on his discretion, her desert.
The anxiety of each revolving Hour
Offers enough for kind participation.
DUNSTAN. Trifles like those a monarch should disdain.
ALDHELM. No, they're the natural cares of natural man.
The rest is all sophisticated wisdom,
Or envious discord rais'd from mad ambition.
DUNSTAN. Heard I aright? speaks Aldhelm thus of marriage?
Of Women? – Nobles! could I hear aright? –
Beseech the holy Bishop to explain
Lest on your mind's remain some strange suspicion.
ALDHELM. Dunstan, I ever thought domestic life
Declin'd by holy monks, midst other blessings
Dear to the soul of man, lest it too fondly
Should cling and fasten to the inward Heart
Which pious pennance ought to call it's own.
If Woman is our scorn, Wedlock, our horrour,
Where dwells the virtue of our self denial?
DUNSTAN. Discussions on this point were dangerous;
Greater and greater grows the threaten'd mischief.
Assert yourselves, ye guardians of the Land,
Nor suffer with impunity such crimes.

Scene v

ODO. DUNSTAN. ALDHELM. SIGEBERT. CERDIC.
ERNULF. REDWALD. LORDS.

Enter EDWY, *attended by* ORGAR.

EDWY. Lords of my council! Nobles! Prelates! Friends!
If ye have souls open to Honour's call,
If ye have spirit to resist foul injury,
Come forth! and aid me at this dreadful juncture! –
If ye have horrour of the blackest treachery,
Deeds of fell baseness! – cruel! – dark! – inhuman! –
Rouse! 'tis your monarch sues you for redress!

III iv 53 *Or envious ... ambition*: an insertion, replacing 'Or mad ambition rais'd from envious discord'.

O if ye have Hearts susceptible of pity
Behold your King its object! – all bereft
Of all he cherish'd most! – disgrac'd – dishonour'd –
Blasted in every Hope! – O rouse! come forth!
Avenge my crying wrongs, Friends! Nobles! Prelates!
ALDHELM. Is there a Breast this anguish does not pierce?
Nobles, the injuries of a virtuous Prince
Are wounds that rankle in a nation's honour
Till every of his subjects, first, and last,
Lordliest, and meanest, rise to do him justice
As Champion's individual. Name the offenders!
EDWY. They are hid in mantles of obscurity
Dark as their crimes; all but the perjur'd Leofric,
Whom flight condemns. Ye, then, who own me King
Oh head your vassals, and pursue the Traitors! –
CERDIC. A call like this might turn an Enemy.
Great Sir, we are yours, to vindicate your rights,
Or, with our lives, to answer for your wrongs.
DUNSTAN. Yes, gracious Sir, your Subjects are your Creatures,
Bound to obey your Will, to court your favour,
Live by your Smiles, or perish by your Frowns
As suits your pleasure.
EDWY. Dunstan, Heaven forfend
I should so lightly weigh the rights of Nature,
Or ballance so unskilfully my own
To think Heaven form'd a nation for my Will!
No; I but claim the justice which I practice,
The common rights of Equity and Law.
DUNSTAN. My Lord, I spoke but of the Seculars:
Not such the sacred body of the Priesthood;
That forms no vassals for mere Earthly monarchs;
'Tis lent to teach, not to be rul'd by kings; –
Nor ever prosper those who slight such precepts.
EDWY. Ha!
DUNSTAN. The Convent's long neglected reformation
Has wraught this ill. Ours are those seminaries;
And to partake them with the Seculars
Is profanation.
ALDHELM. How? which way profane?
DUNSTAN. What! can a Prelate doubt monastic rights?
EDWY. Seize not this anxious Hour for such discussions,

III v 21 *condemns ... King*: an insertion, replacing:
betrays. I've sent to every quarter, –
And here I summon all who own Me King.
III v 22 *Oh head your*: an insertion, replacing 'To head their'.
III v 41 *ill*: an insertion, replacing 'evil'.

But rather –
DUNSTAN. Sire, what time can suit so well?
Thou'rt wretched; trace the cause; 'tis sacrilege. –
Yes! now I comprehend my frequent visions –
Attend, o King and Council! whilst I speak them.
I have been visited by evil Spirits,
Malignant Daemons have assaulted me
By Day, by Night –
EDWY. Dunstan, some other time. –
Lords, to the point and purpose. –
DUNSTAN. Hear me out.
Thyself, and all thy race, and all the Nation,
Will hotly be pursu'd by fierce disorder
Till pardon be obtain'd by due Reform
Of the whole kingdom's Convents. I have spoken!
ALDHELM. List not to this presumptuous egotism,
Nor waste thy faculties on superstition.
Distribute justice with an equal hand
To Church and State, unmov'd by Threats, or Flattery:
The Nation claims at large thy guardian care,
Each party, every class, alike thy subjects.
DUNSTAN. Yes! I foresee the awful punishment
That with an Earthquake's Gulph will swallow us!
The holy function of its dues is robb'd
To bribe the guilty secular's consent.
EDWY. To bribe? most insolent! Dunstan, take heed!
Edred might be thy Dupe, – not so his Nephew! –
DUNSTAN. O wide indeed from this the Days of Edred:
I was his conscience-keeper, Counsellor,
His public agent, and his private guide.
EDWY. Thou wert his agent? true; – I know it well: –
Once more, take heed! –
DUNSTAN. I see, I see Perdition
Arm'd with Destruction's powers, hang hovering o'er us! –
Resign thyself, young King, to admonition
And rigorous pennance, or the Land is lost.
EDWY. Dunstan, the vices of thy early youth
May need monastic terrors for redemption,
Nay, call for them aloud: the youth of Edwy

III v 50 *evil Spirits*: the reviewer for *The Review* objected to this phrase as one 'which we particularly recommend [Burney] to expunge' (22 March 1795).
III v 65 *Earthquake's*: this word was also criticized in *The Review*.

III v 47 *Thou'rt wretched*: an insertion, replacing 'If thou'rt unhappy'.
III v 58 *not*: 'my lord' is deleted.
III v 67 *To bribe ... consent*: an insertion, replacing 'To bribe to criminal consent the seculars'.

Defies the breath of slander to attaint it:
Nor will I mock the truth I venerate
By feign'd contrition of a virtuous passion.
DUNSTAN. Will not the roof drop in? Nobles! submit ye
To this avow'd disdain of one and all?
REDWALD. Disdain? of us? His Nobles? –
DUNSTAN. Redwald, yes;
Without a blush he braves you with his shame.
EDWY. When Edwy blushes, let it be for guilt!
If I offend a Law of Heav'n or Man,
If aught but a mere arbitrary edict
Opposes my alliance with this Princess,
I will resign her! penitent and humble,
Clear her fair fame, – and then, no more behold her!
DUNSTAN. What bold impiety! Will no one speak?
Ruin on ruin falls upon our Heads. –
The papal power arraign'd – its justice scoff'd –
A Courtezan upheld –
EDWY. Dunstan, no more!
Hear first thyself arraign'd! – Arraign Me after. –
Lords of My Council! – In the Face of Day, –
In presence of ye All, – In sight of Heaven –
ALDHELM. O hold, my Prince! –
EDWY. I know what thou wouldst say; –
But 'tis too late! – Leave Prudence for the Guilty! –
I here attest That Dunstan is a Traitor!
DUNSTAN. Me? –
ODO. Dunstan? –
ERNULF. The good Abbot? –
REDWALD. How? a Traitor?
EDWY. I'm prob'd too deeply! – Yes! a Traitor, Lords! –
If not, come forth, Abbot of Glastenbury,
And answer to My charge. What hast thou done
With the vast hoards of unaccounted Treasure
Of which, in the late reign, thou drain'dst the Coffers,
With grasp rapacious, of My credulous Uncle?
DUNSTAN. O Times of most degenerate sinfulness!
The pious Edred, Sir, disdain'd to hold
His Confessor, his guide, responsible
For dross so vile as circulating Wealth:

III v 84 *Will not the roof drop in?*: *The Observer* reviewer declared that 'Dunstan excited the risible faculties of the audience, by some strange ejaculation about the roof's falling' (22 March 1795).

III v 82 *truth I venerate*: an insertion, replacing 'holiness of piety'.
III v 92 *penitent and humble*: an insertion, replacing 'with the humblest penitence'.
III v 104 *How? a Traitor?*: an insertion, replacing ''Tis not possible!'.

Vainly thou question'st me; vain is each menace.
I scorn reply.
EDWY. And is it thus thou answerest?
DUNSTAN. It is.
EDWY. Hear, then, thy sentence –
DUNSTAN. Sir, beware!
'Tis not for Worldly interests, such as these,
To attack, or judge, Ecclesiastic rights.
Beware, I say! nor let thy lawless passion
For this frail woman madden thee to ruin. –
EDWY. Insulting hypocrite! barb'rous Defamer!
I do believe thee author of her seizure – –
O beauteous Sufferer! whither art thou forc'd? –
Dunstan, receive the mede of thy contempt:
Forthwith, I banish thee from My Dominions. –
– I will not hear a word! – Orgar, be't thine
To see him strictly guarded through My Kingdom,
Here to return no more on pain of death.
ODO. Permit me, Royal Sir, –
EDWY. I will not listen!
Pleas, exhortations, prayers, are all in vain.
The treasures of the State he has embezzled –
O what of those to one sweet Gem purloin'd! –
Guards! take him hence! – My Lords, all suit is vain.
Away with him! – I would be left – stay, – Aldhelm!

DUNSTAN *walks sullenly off with* ORGAR, *surrounded by* GUARDS. *The Rest retire precipitately.*

Scene vi

EDWY. ALDHELM.

EDWY. I see thy brow severe; I know thou blam'st Me: –
I see, too, Sorrow sits upon that brow

III v 115–16 *Vainly thou ... reply*: an insertion, replacing:
Nor ever stoop'd I, in those holy Days,
To care, or thought, of what became of it.
Base, sordid mineral! What! am I its keeper?
III v 116 *And*: an insertion, replacing 'Specious evasion!'.
III v 122 *barb'rous Defamer*: an insertion, replacing 'Inhuman libeller'.
III v 134–35 *My Lords ... Aldhelm!*: an insertion, replacing:
He has insulted Me. –
Away with him! – No applications, Nobles! –
I would be left by all. – Or all but Aldhelm.

Mingling with censure pity; – spare me both!
I can endure nor wrath, nor tenderness,
Melting, nor irritation. –

ALDHELM. O my Lord!
If on this brow, in characters of truth
Is writ what passes here – Thou read'st not solely
Feeble compassion, nor unkind displeasure;
Thou read'st, my Leige, a Soul of sympathy
Fill'd with thy wrongs, and all on fire to serve thee.

EDWY. Call, then, a synod to confirm My marriage,
And clear – O though I lose her! – clear the fame
Of My chaste, injur'd Wife! – Yes; let her rather
Conceive Me alter'd, false, ingrate, – than doubt
That her fair Honour was by Me deluded, –
Me, whom she trusted, – Me, whose faith was pledg'd
More than My life – more than Her Love to guard it!

ALDHELM. Weigh first, my Lord, how dangerous at this crisis
To summon the offended Ecclesiastics. –

EDWY. Nay, reason not; – obey Me! – or be silent! –

ALDHELM. Alas, I fear Divorce, not confirmation
Will at this perilous juncture –

EDWY. How? Divorce?
They will not – dare not! – Aldhelm, call the Synod!
Command it's justice! – Nay, nay, speak no more!
Sigebert!

Exit ALDHELM.

Scene vii

EDWY. SIGEBERT.

EDWY. What News?

SIGEBERT. None, sir.

EDWY. Impossible! –
Where hast thou been? Which way? Speak! Answer quick! –

SIGEBERT. All round the Palace parties are dispatch'd.

III vi 11 *a synod*: an assembly of the monastic clergy.
III vii 1 *None, sir*: *The Observer* reviewer claims that Sigebert here replies 'sheepishly' and then 'retires, as if he had forgot what brought him forward'. In the text, however, there is no indication that Sigebert should leave the stage.

III vii 2 *quick*: two lines are deleted:
East? – West? – North? – South? – Dost thou not know thy course? –
Tell me at once. –
SIGEBERT. My Lord, by each direction

My orders were to wait and watch within.
EDWY. She's lost! and I shall never see her more! –
This coldness of pursuit – Sigebert, by Heaven
Thou couldst have found, have rescu'd, have restor'd her,
If earnest, if –
ELGIVA (*within*). O let me pass! –
EDWY. High Heaven! –

Scene viii

EDWY. SIGEBERT.

ELGIVA *rushes in*, ELTRUDA *follows*.

ELGIVA. Where is my Lord? – O where? –
EDWY. My Elgiva! –
O all ye Powers of Mercy, 'tis herself! –
Sweet pitying Angels! 'tis my Elgiva! –
ELGIVA. Take, take me yet again! – My Lord! my King!
My Husband! – O once more –
EDWY. Once and for-ever!
Yet, yet I scarce believe –
ELGIVA. Ah! they pursue me –
I hear their footsteps! – Hark! – O save me, Edwy! –
EDWY. Whom dost thou fear? My love, my Elgiva,
Whom fear'st thou?
ELGIVA. O! – they mean to murder me! –
I know it well! Ah, shield me –
EDWY. Shield thee? Yes!
My life is now entwin'd around thy safety,
And through the medium of their Sovereign's Heart
Alone shall this sweet fabric be assail'd.
ELGIVA. O grateful sounds! O soothing to mine Ear!
Can I again be thine? My Lord! my Husband!
Where are my sorrows flown? This happy Heart
Already spurns them all! forgets, forgives,
Revives – Ah Heaven! what noise – what sound is that?
Should it be Dunstan –
EDWY. Dunstan?
ELGIVA. O my lov'd Lord!

III vii 7 *have rescu'd ... her*: an insertion, replacing 'her twenty times ere this'.
III viii 5 *for-ever*: 'Elgiv! my Elgiv! – Extacy of joy! – ' is deleted.
III viii 9 *Whom fear'st thou*: an insertion, replacing 'Am I not with thee?'.

Let him not tear me hence again! O save me!
O rather let me die! –

EDWY. Inhuman Villain!
Was't even so? Eltruda, speak!

ELTRUDA. Yes, Dunstan.
And to the Irish shore he meant to send us;
But wonderful was our escape: a soldier
Into whose hands, by happy chance, I fell,
My foster-Brother prov'd; and though by Dunstan
Instructed 'twas Religion to destroy us,
Remembrance, gratitude, and quick surprise,
With promise of reward –

EDWY. Reward? My Kingdom
Were insufficient for the noble service. –

ELTRUDA. He rescued us, –

EDWY. Unfold the rest at leisure.
What hast thou suffer'd, loveliest, gentlest, dearest!
Thy agitated brow still speaks thy terror. –

ELGIVA. 'Tis but it's shadow! all is peace within:
There grateful joy, there, pure felicity
Resume possession in my Edwy's name.

EDWY. My matchless Wife! –

ELTRUDA. Should she not now repose?
Spent and exhausted, – to her own apartment –

ELGIVA. Ah! thence, perhaps, again – My Lord! My Edwy!
I die with fear! –

EDWY. Compose thy trembling Nerves,
We part no more! – our marriage is avow'd:
Claim thy just rights of England, and of Me,
Come to thy own Domain, My Wife! My Queen!

EDWY *leads her out.*

ACT IV

A Gothic Hall.

Scene i

SIGEBERT.

SIGEBERT. Kenred ere this has found the king. Who's there? –

Enter CERDIC.

CERDIC. Confusion and distress meet me each way.
Where is the King? – knowst thou of Elgiva? –
Rumour reports the Princess seiz'd again.
SIGEBERT. Rumour for once is true; scarce was she hail'd
Queen of the Land, by Edwy's royal order,
When the dire sentence of divorce, pronounc'd
By Odo in full Synod, reach'd the Palace.
CERDIC. Dunstan, then, ev'n though banish'd, has prevail'd.
SIGEBERT. Instant the King, on pain of Death, commanded
No one should tell the Princess the Decree:
Then, all unguarded, hastily he flew
In person to assemble a new Synod,
With the fair hope the act might be repeal'd
Ere she could hear it. Rash, successless measure!
No sooner was he gone, than a fierce troop,
By Odo instigated, Ernulf led,
The portals forc'd, and carried off the Princess.
CERDIC. Had I been here! – What! was there no resistance?
SIGEBERT. All that surprise could muster; but 'twas vain;
The arm'd and daring Band bore all before it,
Loudly proclaiming 'twas in Dunstan's Cause.
CERDIC. That hasty exile has involv'd this ruin.
Is Patience never to be found with Love?
SIGEBERT. Eltruda, faithful, to her Princess clung,

IV i 1 *there*: a line is deleted:
Cerdic? –

Enter CERDIC

CERDIC. The same. May I not enter?
SIGEBERT. Yes.

Defying aught but Death to part them. – Cerdic,
'Twas horrible! – the cries of Elgiva,
Torn from her home and husband, rent the Heavens.
Who could have view'd unpitying her despair?
Queen for a few short Hours, now Queen no more!
Scarce brought to see the Throne she meant to fill,
Scarce spar'd to look on what she thought to hold!
CERDIC. Unhappy Princess! yet more wretched Edwy!
How will he fire, indignant, at this insult!
How will his state, henceforth, grow hateful to him!

Scene ii

SIGEBERT. CERDIC.

Enter KENRED.

SIGEBERT. Kenred! – knows yet the King the fatal news?
KENRED. Ev'n now; and dreadful is his rage, – his grief!
His Guards are order'd out: he means to head them,
And, sword in hand, re-claim his rightful Wife.
He has sent me onward to prepare –
EDWY (*within*). What hoa! –
Summon me Aldhelm! – Quick dispatch for Odo!
At peril of his mitre speed him hither.

Scene iii

SIGEBERT. CERDIC. KENRED.

Enter EDWY. OFFICERS. ATTENDANTS.

EDWY. Destruction light on me if I forgive it!
O be my portion bitterness and sorrow
If e'er – Hah! Cerdic! – I adjure thy aid! –
I think thou art my friend? –
CERDIC. Prove me, my lord!
EDWY. Haste, then, and head my Guards, – March to the West, –
That way the violater's took. – Myself
Will quickly follow. – Fly! – Each moment teems
With added danger. –
CERDIC. Sir, I go; – yet Redwald –
Thus to supplant him –

EDWY. Speak not now of forms;
Some other charge be his. 'Tis My command
That this be thine: – oh! – if thou holdst sacred
Fair Justice, – upright Honour, – a wrong'd Sovereign,
Rescue Thy Queen and Mine! be fleet, and faithful!
CERDIC. My King and Master – with my life I'll serve Thee! –

Exit CERDIC.

EDWY. I now can better wait. – Where is this Odo?
How slack is apathy! Go! bid him hasten!

Exit an OFFICER.

Speed thee to Redwald. Tell him 'tis My Will
Cerdic this once should head his troop.

Exit another OFFICER.

Hah! Sigebert? –
Dar'st thou encounter me? Avaunt! –
SIGEBERT. My Lord? –
EDWY. 'Twas in thy charge I left my Kingdom's treasure –
SIGEBERT. My gracious Leige –
EDWY. I know thy poor excuse;
Thou'llt plead superior numbers? What of those?
True English courage needs no spur but Honour
To brave an Host of foes unaw'd by Justice.
SIGEBERT. Whatever could be done –
EDWY. Did she resist? –
Speak but of Her. Did she –
SIGEBERT. My lord, her cries –
EDWY. Her cries? O piercing sound! – the cries of Elgiva? –
And could ye hear them, yet not die to save her?
Could not that voice wake pity? bid compunction
Steal down the roughest cheek, in huge big drops
Unfelt before? Would not ev'n beasts of prey,
Suspended, midst the pangs of howling Famine
Have listen'd to such accents? – Aldhelm? Is't thee? –

IV iii 28 *her*: two lines are deleted:
Have ye no Hearts? Are ye all form'd of Iron?
Could ye keep life yourselves, yet hear her cries?

Scene iv

SIGEBERT. KENRED. EDWY. ATTENDANTS.

Enter ALDHELM.

EDWY. O! thou art just in time. To thee I leave
This tardy Odo. Tell him, in My name,
If he revoke not his inhuman sentence
I here remove him from all holy functions.
And now, I am gone. –
ALDHELM. What means my lord?
EDWY. I fly
To vindicate My Wife! My Fame! My Honour! –
ALDHELM. Hear me but first! –
EDWY. The World should not detain me. –
ALDHELM. Go, then, my lord. – unhappy Elgiva!
Who may protect thee now?
EDWY (*returning*). Hah? – What sayst thou? –
ALDHELM. Lost if recover'd, hapless ev'n if sav'd –
EDWY. How lost? – How hapless? – Aldhelm quick explain!
ALDHELM. Her Fame demands the penalty of parting
Till this dread interdiction be repeal'd.
If Her thou lovst, then stay! thyself? – why go!
EDWY. Her Fame? – thou bringst me back! – thou tearst my Soul!
Her Fame is dearer than my Crown, – my Life, –
ALDHELM. Bear at this awful crisis then, her absence.
And safe, yet distant, guard her, till the Pope
Reverse the sentence of Divorce that parts you:
So only can her name be held unsullied.
EDWY. Then will I not pursue her, though I forfeit
For ever her lov'd sight! – 'Tis done! – I yield. –
Kenred alone shall follow Cerdic, (*to* KENRED) – Haste!
Enjoin his strictest search; – tell him all hope
Hangs now on him alone; – for Me, – unhappy! –
I dare no longer aid – away! away! –

Exit KENRED.

IV iv 10 *ev'n if sav'd*: an insertion, replacing 'if restor'd'.
IV iv 13 *repeal'd*: an insertion, replacing 'remov'd'.
IV iv 16 *Life*: three lines are deleted, and two revised:
ALDHELM. Wouldst thou restore it spotless as thou found'st it?
EDWY. Would I? Yes! with my blood! –
ALDHELM. Prove it, my lord,
And at this awful crisis bear her absence.
Send those thou best canst trust to rescue her,
And guard her safe, yet distant, till the Pope

Should he o'ertake – Nay, without stopping, hear me! –
follows KENRED *out.*

Scene v

ALDHELM. SIGEBERT.

ALDHELM. Virtuous young Prince! – Could I relieve thy woes! –
Forbearance such as this, suffering, yet willing,
Might melt the flintiest Heart.
SIGEBERT. 'Tis true; and yet
His hasty Temper –
ALDHELM. Youth, and early power,
Sigebert, might make thy Temper hasty too.
Had Years matur'd him ere he reach'd the Throne
His virtues might have rank'd him first of kings.
Form'd mildly to command, or nobly yield,
Fair his intent, his very wishes upright,
His Soul the genuine seat of native honour;
But of a sensibility so ardent
His spirit impetuous owns no curb but feeling:
Reason it leaves behind, impell'd by impulse;
'Tis fiery, ev'n to fury, if incens'd,
'Tis tortur'd, ev'n to madness, if unhappy. –
SIGEBERT. The Archbishop comes. –
ALDHELM. I'll to the king myself,
And strive to brace him for the impending storm.
Exit.

Scene vi

SIGEBERT.

Enter ODO.

SIGEBERT. Thou bringst, I hope, some comfort to the King?
ODO. I bring him his impiety's reward.
SIGEBERT. Is he not punish'd with extremest rigour?
ODO. What could he hope? he weds his Concubine,
And banishes his Confessor. Such deeds –

Scene v: an inserted scene number.
vi: an insertion, replacing 'v'.

Scene vii

ODO. SIGEBERT.

Enter EDWY.

EDWY (*in entering*). Her cries! – they tear my Heart! – they pierce my Brain!
Her cries! – a voice so soft – (*to* ODO) At length, thou art come?
Hear me, and mark! Thou knowst – perchance too well! –
An execrable gang, in Dunstan's name,
Has, in my absence, carried off My Wife,
Thy Queen? –
ODO. My Lord –
EDWY. Know you, abet you, – tell me!
The villainous act?
ODO. Sir, I conclude it done
In pious vengeance for the banish'd Martyr.
EDWY. Pious? O monstrous! – Martyr? – most profane! –
ODO. The people call aloud on sainted Dunstan –
EDWY. I, too, aloud call for my Elgiva!
Shall they be louder, fiercer, more indignant
Than a wrong'd Husband? than an injur'd Monarch?
If I have ta'en Their Idol, where is Mine?
If They can feel, what think they I endure?
ODO. Canst thou, then, Sir, compare –
EDWY. Far be it from me!
Her virtues would outweigh an host of Dunstans.
A single quality of her fair mind
Has more of truth, worth, strength, and energy
Than the joint thoughts and actions of his life.
ODO. I come, Sir, with the Synod's new decree
And with its menace –
EDWY. Menace? And to whom?

vii: an insertion, replacing 'vi'.
IV vii 6 *Queen*: six lines are deleted, and one revised:
ODO. Sir, hadst thou better chose, this evil
Had never happen'd.
EDWY. Better? Intolerable!
What wouldst thou have me wed? a public pageant?
Set up on high to feast the gazing Vulgar?
Without consulting Heart, or Head, or aught
Save sordid wealth, or interest of alliance? –
Of this no more. – know ye, abet ye, – tell me!
IV vii 19 *worth ... energy*: an insertion, replacing 'of worth, of strength, of energy'.

ODO. First, Sir, I would employ fair argument
To induce thee to give up those false affections
Which Reason and Religion both arraign. –
EDWY. Thinkst thou I'll hear the vapid subtleties
Thou durst impose for Reason and Religion?
Born to be sensitive, say, must we suffer
Each human ill, yet spurn each human feeling
For arts of vain presumptuous sophistry,
That, with no standard or for right or wrong,
Cavils to err, and errs to cavil on?
When thou canst fashion to thy Will thy limbs
Then talk of modelling the Soul at pleasure.
True Piety repells all mask, all mockery.
ODO. Thou wilt not hear me?
EDWY. No!
ODO. Be then inform'd
The Princess Elgiva –
EDWY. Hah! – name her not!
Let not a sound that breathes but purity
Mix with this noxious air; all foul with discord.
ODO. Sir, I must speak, lest Excommunication
Sudden involve –
EDWY. How? – Excommunication? –
What she? – What Elgiva? – whose piety
Might turn the Hearts of Infidels to Truth? –
What dost thou say? – Forbid it, Heaven and Earth! –
ODO. My Lord –
EDWY. Thou hast struck me to the Heart! – that word –
O save her, Odo! – I adjure – I charge thee!
Nay, beg, – nay humbly sue – save her that horrour! –
ODO. Plead for thyself, my lord; – thou still hast time;
For her – the sentence is already past.
EDWY. Thou durst not say'it! – Deny! – revoke! – retract – !
Call back the Edict! – Blot it from the World! –
Be quick! – I breath not till 'tis done! – Away! –
Why dost thou stand?
ODO. My lord, there's no appeal
Till thou concurr'st the marriage be annull'd.

IV vii 23 *would*: an insertion, replacing 'am willing to'.
IV vii 24 *false*: an insertion, replacing 'vain'.
IV vii 26 *hear*: an insertion, replacing 'listen to'.
IV vii 27 *Religion*: three lines are deleted, and one revised:
Were we not born to feel, ere taught to argue?
Who was it made us sensitive? the same
That made us hear, see, smell! gave us each faculty,
And form'd us what we are.
IV vii 30 *For arts of vain*: an insertion, replacing 'For the vain efforts of'.
IV vii 38 *but*: an insertion, replacing 'the breath of'.

EDWY. O black alternative! – Monsters of cruelty! –
Must I then lose her, or myself condemn her?
Savage device! – My swiftest Horse! –

Exit an ATTENDANT.

ODO. My lord
The moment thou pursu'st, thou shar'st her doom.
EDWY. Unheard barbarity! Must I conspire
To abandon all I should protect, or risk
Confirming, by partaking the anathema? –
I'll speak with thee no more! – I feel on fire! –
I dare not trust myself! – Wait here My answer.
Aldhelm shall bring to thee My last resolve.

Exit.

Scene viii

ODO. SIGEBERT.

SIGEBERT. Unhappy King! Who would not pity him?
ODO. Full soon will penitence o'ertake his rage.
The impious exile of the holy Dunstan
Will fall on his own head.

Scene ix

ODO. SIGEBERT.

Enter ORGAR.

ORGAR. Speed to the King,
Sigebert, and let him know the people rise
In clamourous multitudes, and call for Dunstan.

Exit SIGEBERT.

Scene viii: an inserted scene number.
IV viii 4 *head*: three lines are deleted:
That Prince of Monks
Told me, in parting, many nightly visions
Of evil Spirits, which had set afloat
This mischief in the state.
ix: an insertion, replacing 'vii'.

Scene x

ODO. ORGAR.

ODO. And they do well. He told me, as we parted,
His nightly visions of the evil Spirits
Which set afloat this mischief in the State.

Scene xi

ODO. ORGAR.

Enter ALDHELM.

ALDHELM. I bring thee, Odo, the King's final wishes.
Ah! let them not prove vain! Still sits the synod?
ODO. Yes; and awaits my embassy, to fix
If Elgiva alone is excommunicate. –
ALDHELM. Ill-fated Princess! By command o' th' King
I go myself, his messenger, to Rome
To lay his cause before the holy Pontiff.
Suspend then all Decrees – He asks no more,
Till I make known the doom irrevocable.
ODO. If Dunstan were recall'd – –

Scene xii

ODO. ORGAR. ALDHELM.

Enter EDWY.

EDWY. Still art thou here?

Scene x: an inserted scene number.
IV x 1–3 *He told me . . . State*: these lines are a revision of Odo's deleted lines in Scene viii. They replace three deleted lines here:

They follow him in troops.
Wisely he said his banishment would prove
The signal of a War 'twixt king and Monks.

xi: an insertion, replacing 'viii'.
IV xi 9 *I*: an insertion, replacing 'my return'.
xii: an insertion, replacing 'ix'.

Still must mine Eye meet thy pernicious aspect?
Haste to the Synod; bring Me its resolve;
And, patient as I may, I'll wait thy answer.
ODO. Permit –
EDWY. I pray no more!

Exit ODO.

Scene xiii

EDWY. ALDHELM. ORGAR.

EDWY. No news from Cerdic?
How shall I pass this torturing length of time?
(*to* ORGAR *who goes out*).

Scene xiv

EDWY. ALDHELM.

Enter SIGEBERT.

SIGEBERT. Sir, Redwald Earl of Mercia craves your leave
He may resign.
EDWY. O, tell him he's most welcome!
A partizan of Dunstan! – If he's anger'd
Cerdic has his command, let him be anger'd!
Could I have trusted him, I had not chang'd.
Bear to him this reply.
ALDHELM. Will't not inflame –
EDWY. Let it inflame! Bear to him My reply!
Cerdic deserves the charge.

Exit SIGEBERT.

IV xii 5 *Permit ... more*: an insertion, replacing:
My lord, I know –
EDWY. I hear not what thou knowst!
Bring Me the last result! –
Scene xiii: an inserted scene number.
IV xiii 2 *torturing*: an insertion, replacing 'lingering'.
xiv: an insertion, replacing 'x'.

Scene xv

EDWY. ALDHELM. ATTENDANTS.

EDWY. Aldhelm, I see
Thou thinkst me violent? – Perchance I am! –
Wouldst thou be tamer, suffering wrongs like Mine? –
Thou dost not speak? – Well, – silence may be wiser. –

Scene xvi

EDWY. ALDHELM. ATTENDANTS.

Enter ORGAR.

ORGAR. The people grow outrageous, Sir; the Monks
Excite them to arise. –
EDWY. On what pretence?
ORGAR. The exile, Sir, of Dunstan. –
EDWY. Hateful Name!
On thy allegiance, utter it no more! –

Exit ORGAR.

Scene xvii

EDWY. ALDHELM. ATTENDANTS.

Enter SIGEBERT.

SIGEBERT. The cry for Dunstan, Sir, –
EDWY. What cry? – What Dunstan? –
Dost thou not know my orders? Learn them now:
I will not hear that Miscreant Monk exists!
And he who next obtrudes on Me the sound
Forfeits My favour and My sight for-ever!
Leave Me! –

Exit SIGEBERT.

Scene xv: an inserted scene number.
xvi: an insertion, replacing 'xi'.
IV xvi 4 *more*: two lines are deleted, of which the first is illegible; the second reads 'Bring Me no further news till Odo come'.
xvii: an insertion, replacing 'xii'.

Scene xviii

EDWY. And leave Me All! I am in no state
For Witnesses! – I am spent – distract – Away!
Give me a little peace! – Aldhelm, remain.
Exeunt ATTENDANTS.

Scene xix

EDWY. ALDHELM.

EDWY. How shall I live an Hour in this despair? –
Speak, Aldhelm, speak! – but utter not reproach;
My Brain is all disturb'd, and will not bear it. –
Speak! – but to calm, not irritate my woes. –
Tell me which way to turn, what course persue –
Quick point me out some means to rescue Elgiva –
Speak! – but upbraid me not. – Why art thou silent?
Where is that Wisdom, fertile in expedients?
That Judgement, which set human ills at naught?
That sage Experience, which defied surprise?
That Heart benign, which pitied ev'n in wrath? –
Speak, Aldhelm, speak! – Where are thy faculties?
ALDHELM. Alas, till Reason, the high harbinger
Of Sorrow's only friend on Earth, Religion,
Assert its empire o'er thy troubled mind
With renovating strength to strive 'gainst evil,
Counsel were fruitless. Yet perhaps the future –
EDWY. What of the future? – lengthen'd misery
Is all 'twill bring to me! –
ALDHELM. Despair not, Sir;
Shall time have power to mitigate remorse –
That deadliest malady of human culture! –
Yet have no influence over virtuous sorrow?
EDWY. He knows no sorrow, that can find its cordial. –
ALDHELM. This is the language, Sir, of ardent Youth,
All uninstructed in the common lot,
And in the strength of Man to bear misfortune.
Calamity is mortal, like its sufferers:
But though pure bliss is all above, ev'n here
Our minds are fashion'd in so just a form

Scene xviii: an inserted scene number.
Scene xix: an inserted scene number.
IV xix 13 *Alas*: 'my lord' is deleted.

To suit each change of sublunary being,
That, trust me, Prince, there is no Earthly ill
But has an Earthly cure.
EDWY. A cure?
ALDHELM. Yes; Time.
EDWY. Talk not to Me of Time! – unless thou nam'st it
To execrate its torturing tardiness!
ALDHELM. Yes, Sir, in Time there is a magic charm
All imperceptible, yet all prevailing,
Too fine for observation, and too subtle
For opposition, that yet works such changes
In mortal man –
EDWY. Aldhelm, no more! I know
What those wouldst say, and what infer. Thou wrongst me! –
ALDHELM. My lord?
EDWY. Can Time compose a grief like Mine?
Estrange this faithful Heart from Elgiva?
Thou hast touch'd a string that will not bear vibration!
ALDHELM. My Liege? –
EDWY. Thou knowst Me not! – I pray thee leave Me! –
Thou thinkst Me futile – fickle – worthless – Go! –
Add me not penitence to what I suffer, –
Let me not speak thee ill, howe'er thou movst me! –
ALDHELM. My lord, if Truth –
EDWY. Not Truth, but vile suspicion.
I feel its falsehood thrilling through each nerve.
Thou deemst my love the meteor of the Instant,
Now here, – now lost, – Aldhelm begone! and spare me
The shame, the grief, – the guilt to injure thee!
ALDHELM. Thou canst not, sir! thy fluctuating spirits
Are all at War against thyself, not me.
Would I could calm them with my dearest blood!
Yet, ere I quit thy presence, deign to hear me.
If thou believ'st I meant to wrong thy love
Thou art wide from all I purpos'd. Just was thy choice,
For she was gentle, meek, benign and good;
Her virtues honour'd what her beauty grac'd, –
EDWY. O Aldhelm, dost thou think her thus indeed?
O reverend Father! dost thou own her worth?
O let me fold thee to my bleeding Heart! –
Forgive – forget – the ravings of my woe! –

IV xix 39 *Aldhelm, no more!*: in his notes on the play (see above, p. 6), Alexandre d'Arblay objects that this phrase is repeated too often.

IV xix 38 *yet works such changes*: an insertion, replacing 'works such wond'rous changes'.

ALDHELM. O Edwy! in thy days of prosperous fortune
A frown unjust had banish'd me thy presence:
I had turn'd Anchoret, ere endur'd disdain
From haughty happiness; but now, O Prince!
So forcibly I feel the revolution
In all thy wishes, views, expectancies,
I cannot anger at thy irritation,
Nor murmur, though lament to be its object.
EDWY. To Thee could I pronounce the word begone?
O Aldhelm! O my venerable Friend!
Stay with me ever, keep beside me, watch me; –
Without thee what were Edwy? – O support me,
Lest some wild act of frantic desperation
Make me commit – what now alone remains
To render me more wretched – some strange crime.

A shout without

What noise is that?

Scene xx

EDWY. ALDHELM.

Enter SIGEBERT.

SIGEBERT. Pardon, my liege, this boldness –
But all are up in arms. The tumult spreads
From Class to Class: the Nobles, Monks, and Rabble
Confus'dly mixing, with one common voice
Clamour for Dunstan, and, with ceaseless din
Claim his recall, or . . . menace Civil War! –
EDWY. Arms, then to Arms! – A thousand Civil Wars
Were less pernicious than that Man's recall!
Arms! All to Arms!

Exit, followed by SIGEBERT.

ALDHELM. Alas! my King! my Country!

IV xix 67 *Anchoret*: a recluse, hermit.
IV xx 9 *Alas! my King! my Country:* since the *Observer* reviewer states that Aldhelm 'closes the 4th act, by an unfinished interjection, "Alas! my King, alas"', this is probably one of the many places where Palmer forgot his lines. The reviewer continues: 'had he, like Lord *Burleigh*, in the Critic [Sheridan's parodic comedy of 1779], shook his head only, he would have expressed himself more eloquently'.

IV xix 70 *wishes, views*: an insertion, replacing 'views, thy wishes, thy'.
xx: an insertion, replacing 'xiii'.

ACT V

A Forest.

Scene i

Enter two RUFFIANS, *followed by* DUNSTAN *in disguise.*

1ST RUFFIAN. 'Twas somewhere here. –
2ND RUFFIAN. I know the place full well.
DUNSTAN. But are you sure you struck her to the Heart?
1ST RUFFIAN. Yes, twice.
2ND RUFFIAN. I saw her fall.
1ST RUFFIAN. Her blood gush'd out;
Behold my Hand!
DUNSTAN. I ask not circumstance.
I would forget her now. – You are sure she is dead?
1ST RUFFIAN. Ay, long ere this: but mine be the reward:
He would not strike.
2ND RUFFIAN. She look'd so innocent,
My mind misgave me. But I kept o' th' Watch,
And therefore –
DUNSTAN. You shall both be recompens'd.
1ST RUFFIAN. Both? No; the deed was mine: he gave no help.
2ND RUFFIAN. Had you but seen how innocent she look'd! –
DUNSTAN. Why dwell on that? I ask not of her looks, –
She's gone; – but what became of her companion?
1ST RUFFIAN. I think I struck her too; but am not sure.
Our orders were but to dispatch the Princess
If the King's party, with strong force, o'er-took
The troop that guarded her, This is perform'd.

V i 2–3 *are you sure ... twice*: this resembles the exchange in Shakespeare's *Macbeth* between Macbeth and the First Murderer:
MACBETH. Is he dispatched?
MURDERER. My lord, his throat is cut:
That I did for him. (III.iv. 16–17)
As in *Macbeth*, in which the murderers succeed in killing Banquo but let his son Fleance escape, the ruffians here wound Elgiva fatally but fail to harm her maid Eltruda.

V i 4 *Behold my Hand*: an insertion, replacing 'My Hand is red with it'. 'Behold this Hand' was first inserted, then deleted.
V i 4 *DUNSTAN*: 'So let it be! – ' is deleted.
V i 4 *circumstance*: 'Unhappy Wretch! – ' is deleted.
V i 6 *reward*: 'Alone' is deleted.
V i 17 *This is perform'd*: an insertion, replacing 'that done, we left her'.

Scene ii

Enter ERNULF, *with* SOLDIERS.

ERNULF. Holla, there! Speak! Who are You? Of which side?
DUNSTAN. Ernulf?
ERNULF. Most happy chance! News of thy rescue
Reach'd me but now; and that thy course was bent
This way, to join us in the Vale of Evesham.
DUNSTAN. The rising crowd revok'd my banishment
With holy zeal. My liberty restor'd,
Eager to serve my Cause, I cross'd the Country
And trust my safety with thy valiant Troops.
ERNULF. Thy sight alone will bow them to thy orders:
Thy very name like inspiration fires them:
Yes! thou hast been obey'd! Dunstan, – alas! –
I had nearly said too strictly! –
DUNSTAN. How? does Ernulf
Shrink back, unsteady, from our righteous cause?
ERNULF. Cerdic, commanding Redwald's chosen Guards,
Persu'd th'ill fated Elgiv, Danger press'd fast –
Be satisfied the fated deed is done.

V ii 2–3 *News of thy rescue ... now*: this revision responds in part to a criticism made by d'Arblay, in his notes on the play, that Ernulf would have had to flee from Cerdic's pursuing troops. He could have come back (*'revenir'*) to meet Dunstan, but not merely stayed in the same place where he attacked Elgiva.
V ii 4 *the Vale of Evesham*: a fertile valley in Worcestershire. Evesham, on the right bank of the River Avon, was first established in the eighth century.
V ii 7 *Eager to serve my Cause*: the phrase 'In this disguise, unknown' was removed because, as d'Arblay points out in his notes on the play, Dunstan is still recognizable to the other characters.

V ii 2 *News of thy rescue*: an insertion, replacing:
I came to seek thee.
The news of the revolt which rescu'd thee
V ii 5 *crowd*: an insertion, replacing 'multitude'.
V ii 6–7 *With holy ... Cause*: an insertion, replacing:
And gave me, with an holy zeal, my liberty:
In this disguise, unknown
V ii 8 *And*: an insertion, replacing 'To'.
V ii 11 *thou*: 'and it' is written above the line, but 'thou' is not deleted.
V ii 13 *Shrink back ... our*: an insertion, replacing 'Forsake, or shrink from, such a'.
V ii 14 *ERNULF*: 'Be satisfied. The fatal deed is done!' is deleted.
V ii 15–16 *Persu'd ... done*: five lines are deleted, and one revised:
Persu'd our victim Prize, the ill-fated Elgiva; –
Danger press'd fast, when, as thou bid'st, yon men
Forc'd her apart, and, hard-by, – murther'd her! –
Eltruda, clinging close, has shar'd her fate;
Or soon, perforce, must perish. On we rode
As nothing missing, to deceive all search. –

Yet griev'd was every Heart.
DUNSTAN. Thou hast done thy duty.
The safety of England demanded her blood.
Ere now, the Excommunication's past –
These Eyes shall close no more till Edgar reign! –
Exeunt.

Scene iii

ELGIVA. ELTRUDA.

ELGIVA *enters with a Bandage tied across her Breast, tottering, and leaning upon* ELTRUDA.

ELGIVA. O! I can crawl no further! (*She drops down.*)
ELTRUDA. Here then rest, –
Sweet suffering Pattern of all Earthly patience!
Alas, thou look'st – How art thou?
ELGIVA. Hold me tighter –
A dizziness – My Head, – My Eyes, – a swimming –
Let me lie down I pray, – Be still! – be still! –
ELTRUDA. O I will watch by thee in softest silence; –
Could'st thou repose – Already in a slumber? –
Or are her senses numb'd? – How deep she sighs!
O she will die! – those livid lips – that paleness –
– Is it sweet sleep steals on her loaded Eye-lids? –
Or but fatigue, that conquers even pain?
While grief itself, relinquish'd to despair,
Ceases to struggle with her wearied nature?

V ii 20 *reign*: the extensive deletions at the end of this scene correspond to suggestions made by d'Arblay in his notes on the play.
V iii 1 s.d. *tottering*: perhaps an allusion to the hamstringing of Elgiva recorded by Hume and other historians; see *History*, I, 95.

V ii 18 *The safety ... blood*: an insertion, replacing five lines:
She was but Edgar's Concubine. My visions
Were black and ireful to demand her blood.
Nor will this expiate. Edwy's self must die.
So say the Spirits that visit me by Night, –
Or Dunstan falls, – and with him sinks all England.
Three revisions were made to these lines before they were deleted: 'fall', replacing 'die' (l. 3); 'sinks', replacing 'falls' (l. 5); and 'England's lost', replacing 'all England' (l. 5).
V ii 20 *reign*: three lines are deleted:
I but await the faithful Leofric
For tydings to absolve us from allegiance.
Come, then, and guard me down the Vale to meet him.

Might ev'n this dread tranquility recruit –
She starts! – How short she breathes! O she will die! –
And yet she bleeds no more, – her wounds are clos'd;
That Bandage has suffic'd – No help? – no succour? –
She starts again! – She's paler – paler yet! –
O my lov'd Princess! O my murder'd Queen!
Die not! – desert not thus thy faithful Eltrud! –
What must I do? – She Wakes! – My Heart's dear Princess! –

ELGIVA (*starting*). Spare, spare me yet! Ah! strike not! –

ELTRUDA. She revives! –
Might she but live! – look, look not thus affrighted!
Raise not thy head thus wildly! –

ELGIVA. Where – where am I? –
Some mist obscures my Sight – O come not near me!
Murder me not! – All, all is dark! – Ah hold! –
Let me but once more see my lord! –

ELTRUDA. Sweet Sufferer!
Thou break'st my Heart! – Turn, trust thyself this way!

ELGIVA. But once! – I ask *no more*! – My Eye-balls pain me:
Ah! is it thee, Eltruda? Who besides
Is here?

ELTRUDA. No one.

ELGIVA. No one? – Who wounded me?
Am I not wounded?

ELTRUDA. O! too sure! too sure!

ELGIVA. I scarsely know. – My good Eltruda, hold me, –
I am faint – I feel quite cold – methinks I die! –

ELTRUDA. Say, say not that! utter not words so direful,
Drive not the poor Eltruda to despair!

ELGIVA. I will not. I am better. Now I know thee, –
I see thee plainer. Good and kind Eltruda!
How came we hither? I have no memory left.

ELTRUDA. The dreadful blow stunn'd all thy faculties; –
But if thou art better – be all else forgot.

ELGIVA. I am; – and yet – I feel it will not last –
But do not mourn –

ELTRUDA. O my poor Heart will burst! –

ELGIVA. 'Twill end a life of woe, – yet woe, Eltruda,
Unmixt, I hope, with guilt.

ELTRUDA. Guilt? Spotless Victim!
Oh thou hast been all innocence – all excellence? –
Too good – too sweet – too pure for this bad World! –

V iii 14 *recruit*: be restored to health, reinvigorated.

V iii 15 *die*: 'Her hand is icy cold, – O for some aid! – ' is deleted. A revision was made to the line before it was deleted: 'No aid? – none! none!', replacing 'O for some aid!'.

ELGIVA. I charge thee, Eltrud, grieve not if I go –
Think of my Edwy solely: – O if yet
For one short instant – one kind blessed moment –
These poor dim Eyes – might they again – O Edwy!

Scene iv

ELGIVA. ELTRUDA.

Enter a PEASANT.

ELGIVA. Ah! who comes hither? hide me! hide me! –
ELTRUDA. Fear not;
'Tis but a Country-man. – Stranger, assist us; –
Noble will be thy recompense.
ELGIVA. Ah! is't not
Some new assassin?
ELTRUDA. No, 'tis a kind Peasant.
Sustain, with gentlest touch, this wounded lady. –
PEASANT. Alack! and who has had an Heart so hard
To shed her blood?
ELGIVA. For me, all aid is past: –
Life quick retires; – one only wish remains. –
Could he but take to Edwy a kind word, –
'Twould bless my parting breath! –
ELTRUDA. Do not talk thus; –
O what of woe has Edwy still in store!
ELGIVA. With pitying gentleness reveal my fate:
Let him not wander through the Desart World
In search of her he never more may find.
When all is o'er – Ah! cease those Tears, Eltruda! –
How can I wish to live? – When all is o'er
Seek him – inform him –
ELTRUDA. No, – revive! revive! –
ELGIVA. Bear to him his lost Elgiva's last love,
Tell him – in Death's cold arms – with grateful fondness –
PEASANT. I see some Soldiers; shall I pray their help?
ELGIVA. O no! o take me hence – conceal me – save me –
Expose me not again to murderous swords! –
ELTRUDA. Alas, alas! know'st thou no place of refuge?
PEASANT. Yes, and close by; on t'other side the Hedge.
ELGIVA. Then lend me each an arm. I am wond'rous faint. –

V iii 51 *Edwy*: 'What wilt thou do? How wilt thou bear this stroke' is deleted.
V iv 4 *kind*: an insertion, replacing 'friendly'.

ELTRUDA. O wretched Hour!
PEASANT. 'Tis but a Step away.

Exeunt, ELTRUDA *and the* PEASANT *supporting* ELGIVA.

Scene v

DUNSTAN. ERNULF. *and* SOLDIERS, *meeting* REDWALD.

REDWALD. I sought, and now have found thee. Edwy comes,
Leading his army, fierce, and fell, and vengeful.
DUNSTAN. Odo ere this has spoke the dread anathema,
Which to the dust must humble Edwy's Crown.
Forward to Battle, then, secure of victory.
REDWALD. I need no urging; Edwy has injur'd me.
I will revenge me both on him and Cerdic.
DUNSTAN. Yet one thing more; vengeance is incomplete
If, when a Tyrant falls, no Heir arises: – –
Where is the gallant Edgar?
REDWALD. Near this spot,
Heading a small detachment for the King.
DUNSTAN. Then let us call him forth; with honours due
Proclaim his rights, and give our cause a Chief.

Scene vi

DUNSTAN. ERNULF. REDWALD. SOLDIERS.

Enter a SOLDIER.

SOLDIER. The King, and his whole Army, march this way.
ERNULF. The King?
REDWALD. So soon?

V v 2 *fierce ... vengeful*: an insertion, replacing four lines:
and in armour clad,
The Princess to regain, or all to massacre.
DUNSTAN. Now, then, ye doughty Warriors, is the time
To save at once the Empire and your lives.
A revision was made to these lines before they were deleted: 'slaughter', replacing 'massacre' (l. 2).
V v 3 *anathema*: an illegible line is deleted.
V v 5 *Forward*: an insertion, replacing 'Fiercely'.
V v 7 *Cerdic*: a line is deleted:
At peril of my life.
DUNSTAN. Nobly determin'd!

DUNSTAN. His men will all be spent;
Now is the moment of attack. –
REDWALD. If Edgar –

Scene vii

DUNSTAN. ERNULF. REDWALD. SOLDIERS.

Enter another SOLDIER.

2ND SOLDIER. The Army of the King hardby has halted;
While he, partaking neither rest nor food,
Surveys, with a slight guard, the adjacent country.
DUNSTAN. What if we now surprize him? – Redwald? – Ernulf? –
ERNULF. First I must know how Edgar stands affected.
REDWALD. And I for Cerdic wait, to meet my charge.
DUNSTAN. Mine be it, then, to parley with Prince Edgar,
And urge him to our side. Lead on the way.

Exeunt.

Scene viii

Enter EDWY, *in Armour,* ORGAR, *and two* SOLDIERS.

EDWY (*to a soldier*). Let their repose be short, then bid them hasten.

Exit SOLDIER.

The body wearies where the Soul's at rest. –
I am glad they are not too wretched for fatigue!
Where are we now?
ORGAR. Hard by the Vale of Gloucester.

V viii 4 *the Vale of Gloucester*: in Gloucestershire, about twenty miles southwest of Evesham, at the head of the Severn estuary on the Welsh border. Gloucester, the county town, situated on the left bank of the lower Severn, was founded by the Romans.

V vi 3 *Edgar*: four illegible lines are deleted, two spoken by Redwald and two by Ernulf.
V viii 4 *Gloucester*: six lines are deleted:
But, good my lord, thus wandering from thy Troop
Dost thou not risk thy life?
EDWY. What if I do? –
Is not my life the forfeit of humanity?
And shall I nurse it up a few poor Years
That pass unheeded, slow, yet wonderous sure,
Lest a brief risk should cut it off with glory?
A revision was made to these lines before they were deleted: 'Is't not', replacing 'Is not my life' (l. 3).

Yet fasting thus, my lord, heated, exhausted,
Might not a little rest –
EDWY. Does Elgiv rest?
Canst thou inform me that? Does soft repose
Steal in sweet slumbers through her gentle Frame,
Oblivious of her griefs? Tell me but that,
Then bid me know the luxury of ease.
ORGAR. The Royal Sufferer, Sir, must soon be found. –
EDWY. Found she may be; – but when? – or where? – or how? –
O who can tell! – Perhaps that lovely Form,
Confin'd by villainous force in brutal fetters,
On the damp, murky Ground of some cold Dungeon
Inhales with every Breath a fell disease:
While the dim light, through curved apertures,
Displays a Scene to make e'en darkness enviable! –
ORGAR. She will be rescu'd, Sir, –
EDWY. Perhaps, escap'd,
Trembling she wanders o'er some desart heath,
From man and Beast an hapless fugitive! –
A thousand nameless terrors at each step
Appal, affright her! –
ORGAR. She will sure meet friends –
EDWY. Famish'd, perhaps, O Elgiv! if thou art gone
Blest be thy Spirit! sacred be thy manes! –
ORGAR. Think not, my Lord, the worst. –
EDWY. Are they not ready?
How slow are those who feel not!
ORGAR. Good my liege,
The time is short; the men were greatly harrass'd;
The quickness of the March –
EDWY. Quick, do they call it?

V viii 25 *manes*: 'the spirit, "shade" of a departed person, considered as an object of homage or reverence, or as demanding to be propitiated by vengeance' (*OED*).

V viii 9 *that*: two lines are deleted:
And thou shalt make for me a Bed of Violets,
Call the blyth Birds to warble me to sleep,
V viii 10 *Then*: an insertion, replacing 'And'.
V viii 18 *e'en*: an insertion, replacing 'grim'; 'even' was first inserted, then deleted.
V viii 24 *perhaps*: three lines are deleted:
the Bird's black Hedge she seeks, –
Her snowy Hands, streak'd with his jetty food, –
A prey as innocent to ravenous Man! –
Perhaps, destroy'd –
The 'jetty' (jet-black) food is blackberries.
V viii 27 *those who feel not*: an insertion, replacing 'the unfeeling!'.

Scene ix

EDWY. ORGAR. A SOLDIER.

Enter the PEASANT.

PEASANT. Good Masters, can you tell me where's the King?
EDWY. What would you with him?
PEASANT. Sir, hardby, a lady –
EDWY. What lady? – Who? –
ELTRUDA (*within*). O! 'tis the King himself! –

Scene x

EDWY. ORGAR. SOLDIER. PEASANT.

ELTRUDA *rushes forward; the* PEASANT *disappears*.

ELTRUDA. Sweet Heaven be prais'd! –
EDWY. Where is my Elgiva? –
What hast thou done with her? – Where hast thou left her? –
Why art thou here alone? –
ELTRUDA. Alas, my lord! –
EDWY. Does she not live? – Speak, speak at once! – be quick! –
O weep not, weep not! –
ELTRUDA. My lord –
EDWY. Say but she lives! –
Give but that comfort to my tortur'd Heart, –
Say but she breathes –

V x 7 *breathes*: three lines are deleted:
I ask no more! – Tell me
The vital air is still – kind, kind Eltruda! –
Say but she lives, and half my Kingdom – Ha! –

Scene xi

EDWY. ORGAR. SOLDIER. ELTRUDA.

ELGIVA *appears, leaning upon the* PEASANT.

ELGIVA. Edwy! –
EDWY (*darting forward to meet her*). 'Tis she! – benignant Heaven! – O extacy! –
Do I once more – O agony of bliss! –
She faints! – Rest, rest on me! – She sinks! – O help! –
Eltruda, – worthy Stranger, –
ELTRUDA. Ah, I fear'd
This sudden joy –
EDWY. Heaven's! how she droops! – and yet
She breathes, – how art thou now? – Silent? – quite silent? –
Wilt thou not speak to thy unhappy Edwy? –
Wilt thou not sooth the anguish of his soul
With one kind word? – one look? – one gentle smile? –
ELTRUDA. See, she revives! –
EDWY. O horrour to my sight! –
Is this my Elgiva? – pale, pale and bloodless! –
Is this –
ELGIVA. O Edwy! O my bosom's Lord!
Take in this last adieu –
EDWY. O dreadful sounds!
O worse than silence! – cruel, cruel Elgiva!
Canst thou thus calmly speak so dire a doom?
Where is thy softness for thy once-lov'd Edwy?
Where – where thy Heart? –

V xi 1 s.d. *ELGIVA ... PEASANT*: Thomas Campbell notes that in the stage production, 'this hedge, though supposed to be situated remotely from any dwelling, nevertheless proved to be a very accommodating retreat; for, in a few minutes afterwards, the wounded lady was brought from behind it on an elegant couch, and, after dying in the presence of her husband, was removed once more to the back of the hedge. The solemn accents of the Siddons herself were no match for this ludicrous circumstance, and she was carried off amidst roars of mirth' (*Life of Mrs. Siddons*, p. 267).

V xi 2 *bliss*: an insertion, replacing 'joy'.
V xi 4–5 *Ah ... joy*: an insertion, replacing two lines:
sweetest Princess.
Alas, I fear'd –
PEASANT. Give her a little air. –
EDWY. My Elgiva!
V xi 12 *Is this*: an insertion.
V xi 12 *my*: 'husband and my' is deleted.

ELGIVA. Ah! bear the coming blow! –
Brings it not peace? –
EDWY. What blow? – What dost thou mean? –
ELGIVA. Alas, I share thy pangs! I feel them all! –
'Tis strange what none escape should prove thus awful:
But who unmov'd can meet Eternity?
I have not been, I trust, a guilty Creature,
And Hope's sweet radiance rises to my view; –
Yet – how am I imprest! –
EDWY. Great Heaven! She sinks! –
Close not thy Eyes thus, Elgiv! Oh look up! –
Raise them once more! –
ELTRUDA. Alas! alas! She goes! –
Life ebbs apace –
EDWY. Why, why that barbarous sentence?
On what presumption?
ELTRUDA. See but her Death-like Face –
EDWY. O sight of woe! – Met we for this at last?
My Elgiva! yet once more speak to me!
Give to my joyless soul one soft sensation,
Suffer that gentle Voice –
ELGIVA. My dearest lord!
It's last weak accents shall pronounce thy blessing,
Thy Elgiv's blessing on thy much-lov'd Head! –
Ah! – were I not so faint – my wounds –
EDWY. What wounds?
Does she then fall by murder?
ELTRUDA. Ah! – they bleed! –
EDWY. They bleed? – O Misery! – every hope is flown. –
My dying Wife! – my parting Elgiva! –
My all of Joy and Good! – My Life's sole solace! –
ELGIVA. O Edwy! lend me courage, – not despair! –
EDWY. My Brain turns round – I know not where I am –
Nor whom – nor what – O kind Distraction seize me!
Merciful Madness! –
ELGIVA. Edwy! generous Edwy!
Repress thy sorrow – for my sake – for me –
Spare, spare the fleeting moments that remain. –
EDWY. I will! – I feel my wrong – Canst thou forgive it?
(*kneels by her side*.)
Can thy pure Spirit – see! I am calm! – forgive me?
ELGIVA. Forgive? – dear to my Heart! – my noblest Edwy! –
Ah me, I die! – O call forth all thy firmness –
EDWY. I will! believe me – trust – I'll wound no more,
Expiring Angel! thy seraphic peace! –

V xi 28 *Face*: an insertion, replacing 'look'.

ELGIVA. Ah! – take my last farewell – my tenderest –
Dies.

EDWY. Thy last? –
ELTRUDA. She's gone? – she breaths no more! –
EDWY. Gone, sayst thou? –
Gone? – and thus gently? –
ELTRUDA. Yes, without a groan. –
EDWY. Away! – She lives! – look at that beauteous face –
All unconvuls'd – Can Death wear such an aspect?
Serene – benign – My Elgiva! – My Wife! – –

Scene xii

EDWY. ORGAR. SOLDIER. ELTRUDA. PEASANT.

Enter SIGEBERT.

SIGEBERT. Arise, my lord! away!
Your army is surrounded –
EDWY. To it, then! –
Haste thee, and save it! (*to* ELTRUDA). – Didst thou say no more?
ELTRUDA. Alas, alas!
EDWY. What, dead? What, gone for-ever?
What! – sunder'd from these arms for life's sad remnant?
What! murder'd, say'st thou? – Tell me the accursd hand – *starting up.*

Scene xiii

EDWY. ORGAR. ELTRUDA. PEASANT. SIGEBERT. SOLDIERS.

Enter an OFFICER.

OFFICER. Cerdic of Mercia is return'd, and craves
Your royal order to attack the Rebels.

V xi 55 *Away*: an insertion, replacing 'Tis false!'.
V xi 57 *Wife*: two lines are deleted:
She does not answer me! –
ELTRUDA. Oh my sweet Princess, –
Art thou no more? –
In their place is inserted, apparently in error, an exchange from scene xiv below between Edwy and Sigebert, who is not on stage:
SIGEBERT. Edgar has join'd the foe
EDWY. Edgar my Brother?
Thou call'st me back to life? I'm wond'rous glad!

EDWY. Cerdic? – He went to seek my Elgiva –
ORGAR. He's here himself.

Scene xiv

EDWY. ORGAR. ELTRUDA. PEASANT. SIGEBERT. SOLDIERS.

Enter CERDIC.

EDWY. Ha! Cerdic! – Hast thou found her?
CERDIC. An horrid rumour met me by the way
And told me – vain the search! –
EDWY. It told thee false!
Lo, where she lies! –
CERDIC. O fierce and barbarous act! –
How shall I speak yet more unnatural horrours? –
SIGEBERT. Edgar has join'd the foe! –
EDWY. Edgar? – My Brother? –
Thou call'st me back to life! – I'm wondrous glad! –
Things the most savage suit my humour best. –
Edgar, where art thou? – Glorious scenes of bloodshed
Fire with new energy my torpid mind. –
going.

CERDIC. Go not alone, my leige;
I'll summon strait the Troops –
EDWY. No, let them follow. –
My Elgiva, adieu! – I haste to join thee. –
going.

CERDIC. Yet hear me, Sir! – Your life –
EDWY. What if I lose it?
Does the whole Earth contain another Being
So willing to be gone?
CERDIC. Yet Head your Army;
Revenge with conquest then –
EDWY. And if I conquer
What gentle bosom greets my victory?
Look there – and answer me? – Will Elgiva

V xiv 2 *met*: an insertion, replacing 'stopt'.
V xiv 6 *SIGEBERT*: an insertion; the line was originally given to Cerdic. This is the exchange also inserted, apparently in error, at the end of scene xi above.
V xiv 7 *glad*: 'My Brother? – Nay, I'll instant then to battle. – ' is deleted.
V xiv 10 *mind*: 'Edgar, thou art lost!' is deleted.

Come forth, with beaming smiles, and Eyes of rapture
To tell me why I fought? – and why I live? –
Away! away! – Edgar! I fly to meet thee!

Exit, followed by CERDIC, ORGAR, SIGEBERT, *and* SOLDIERS.

Scene xv

ELTRUDA. THE PEASANT.

ELTRUDA. O seek some help to guard this honoured Corse

PEASANT. In sooth I'll do my best. I have Friends at hand

Exit.

ELTRUDA. Oh what a sight of woe! break my Heart! –

Re-enter the PEASANT *with two* COUNTRYMEN.

Aid me to move her from this public Path.
Behind these Trees – Ah Heaven! – is't Dunstan Comes
Gently good Peasants – o my Queen! my Queen!

Exeunt, bearing away ELGIVA.

V xiv 21 s.d. *CERDIC*: an insertion.
Scene xv: an inserted scene number. The scene replaces a longer exchange between Cerdic, Eltruda, and the Peasant:

CERDIC. Good Peasant stay; protect that honor'd corse,
And guard this hapless lady.

Exit.

PEASANT. Ay, in good sooth,
I'll do my best. But, lady, there I see
Two men who look like Spies.

ELTRUDA. Ah! Dunstan comes! –
My murder then is next! –

PEASANT. Turn to this side
Till they both pass.

ELTRUDA. What! leave my lifeless Princess?
No, – rather risk –

PEASANT. I'll watch they touch her not.
Come, lady, come; they have left thee to my care,
I cannot see thee kill'd.

ELTRUDA. O my lost Princess! –

The PEASANT *draws her away.*

V xv 2 s.d. *Exit*: an insertion, replacing 'Goes out and re-enters with two other Peasants'.

Scene xvi

Enter DUNSTAN *and* LEOFRIC.

DUNSTAN. This way seems least infested. Leofric
Thy Services are mark'd.
LEOFRIC. The news I brought
Of th'Excommunication of rash Edwy
Through-out work'd wonders. Redwald's party rose
Impatient for the charge; Ernulf's gave o'er
Each ling'ring scruple, and from Rank to Rank
Enthusiasm spread. Prince Edgar's function
Had without Battle clos'd the whole, if Cerdic
Fresh with his haughty troop –

Scene xvii

DUNSTAN. LEOFRIC.

Enter ERNULF.

ERNULF. Edwy is slain! –
DUNSTAN. How? – when? by whom? –
ERNULF. This instant, and by many.
In mad defiance, forth he rush'd alone
And gave the bloody victory at a stroke. –
Most pious Dunstan, – pity more than blame me –
But my Heart smote me as the monarch fell,
And from the Sight I flew – to Thee for comfort.
DUNSTAN. I never meant he should be slain; – confin'd,
Dethron'd, –

xvi: an insertion, replacing 'xv'. 'ELGIVA *dead*' is deleted before 'Enter DUNSTAN'.
V xvi 2 *Thy ... mark'd*: an insertion, replacing 'Thou'lt amply be repaid'.
V xvi 6–8 *Each ling'ring ... whole*: an insertion, replacing:
Each loyal scruple; and, on either side
Rebellion spread apace. Edgar's revolt
Had clos'd the whole without a blow,
xvii: an insertion, replacing 'xvi'.
V xvii 9 *Dethron'd*: 'no more! – ' is deleted.

ERNULF. No more? – Perhaps he may be sav'd –

Exit.

LEOFRIC. How, Dunstan, was't not Thee –
DUNSTAN. Haste and assist –
Nay, – Go –

Exit LEOFRIC.

Scene xviii

DUNSTAN *alone.*

DUNSTAN. 'Tis o'er. I'm glad. – Yes, yes. – And yet
Would he had died a natural Death! – slain! – murther'd! –
Well! – be it so! – 'twas fitting to remove him –
Yet – I feel petrified! – My King! – rash Youth,
Why would he thus provoke – What are these men?
They shrink – they know – or fear me – Hah! a Corpse
Perhaps 'tis Elgiva – yes 'tis ev'n So!
Her lifeless frame – that deed is surely done.
True, as the Villain said, her look is innocent –
Would I had not encounter'd it! – a sickness
Deadly, unfelt before, benumbs, confounds me –
Where may she be? – Who sent her hence? – Was't I? –
By what authority? – Hush! Enquiry! – Hah! –
'Tis Leofric.

V xvii 9–11 *Exit ... Go*: an insertion, replacing:
I fly, I fly! –

Exit.

DUNSTAN. Go thou, too, Leofric.
Scene xviii: an inserted scene number.
V xviii 1 *'Tis o'er ... yet*: an insertion.
V xviii 5–8 *What are ... done*: an insertion, replacing:
hah! who lies here? –
The Form is Elgiva – yes – 'tis herself! –
Her lifeless corse. – That deed is surely done. –

Bends over her.
V xviii 14 *'Tis Leofric*: an insertion. The ms. reads 'Leo'.

Scene xix

DUNSTAN.

Enter LEOFRIC.

LEOFRIC. All has succeeded; naught remains but triumph.
Thy regal foe, bleeding to Death, and fainting,
Is borne to Ernulf's Tent to breathe his last.
Delug'd in Tears, the Victors stand around
And call on Dunstan to appease their feelings,
And satisfy their consciences. For me,
By thee instructed –
DUNSTAN. Leave me, now, a moment
To meditation. Tell them I will come!

Exit LEOFRIC.

Scene xx

DUNSTAN *alone*.

DUNSTAN. With what fell havock does the fiend Remorse
Grasp my disorder'd Soul! cling to my vitals,
Is this my promis'd Joy? O glaze fallacious! –
The World may bow before, and Nation's honour me,
Misguided Myriads may adore my name; –
What's all to him who honours not himself? –
I would exchange me with the veriest Slave;
Lie on the Bed of torture; court the gripe
Of gnawing Famine; beg Bread – and be denied it –

V xx 3 *glaze*: a bright, glossy appearance.

---

xix: an insertion, replacing 'xvii'.
V xix 7 *instructed*: an illegible word or words is deleted.
Scene xx: an inserted scene number.
V xx 2 *vitals*: six lines are deleted:
And, with its forked fangs, wrest all within
From Harmony and order! – Triumph, said he? –
Alas! success and triumph are not one! –
Edwy! methinks thy Sighs, thy dying groans
Sound hallow in my Ears! – shrill shrieks of Elgiva
Pierce me where e'er I turn – away, foul fancy! –
V xx 3 *fallacious*: three illegible lines are deleted.
V xx 4 *The*: '‹whole›' is deleted.

So I were innocent of the blood of Elgiv,
The crying wrongs of Edwy! – Innocent They,
Therefore no sufferers: them, I pity not;
O could I say I do not envy them!
Their task perform'd, their Earthly labour o'er,
Their lasting recompence assign'd in Heaven! –
Where may be mine? – Hah! – bring they Edwy hither?
O let me haste to hide me from the World
Till I can mask the misery of my Soul!

Exit.

Scene xxi

EDWY, *wounded, enters supported between* ORGAR *and* SIGEBERT: ERNULF *and* SOLDIERS *attending*.

EDWY. No, – higher up, – beneath an oak – that way, –
Ernulf, this one kind boon allow'd, exculpates
A world of injury. –
ERNULF. Sir, Heaven's my witness
Your life I sought not. –
EDWY. Spare me now excuses:
I have but time to view the clay cold Corse –
I sink! – Is't borne away? – Stand all apart
Ah! – there she is! –
ERNULF. You hasten, Sir, your end, –
EDWY. No, I but blunt its misery. – Raise – support me
Yes 'tis herself – My Spotless Elgiva –
O higher lift my head! – no more I see thee
With agonizing pain, – no more my fondness

V xx 14 *task perform'd*: an insertion, replacing 'work all done'.
V xx 16 *Hah ... hither*: an insertion, replacing:
In Hell? – horrible contrast! –
– Hah! – bring they Edwy hither? – Is't to probe me? –
V xx 17 *haste to*: an insertion, replacing 'go and'.
xxi: an insertion, replacing 'xviii'.
V xxi 5–6 *the clay ... apart*: an insertion, replacing:
– before I die –
That clay-cold Corse: – nearer yet – take me nearer –
After 'I sink', '*drops down*' was first inserted, then deleted.
V xxi 8–10 *Raise ... my head*: an insertion, replacing:
My Elgiva!

lying down by her side.

Look but what sweetness still sits on that face!
My spotless Elgiva!

Is rack'd with torturing fear; thou art but gone
Where humbly I may hope to follow, – gone
To reap the harvest of thy fair desert.

Scene xxii

EDWY. ORGAR. SIGEBERT. ERNULF. SOLDIERS.

ELTRUDA *rushes in, with the* PEASANT.

ELTRUDA. O my dread liege! The murderers that I fled
Have they then fall'n on Thee?
EDWY. Hapless Eltruda! –
Commend her, Sigebert, to my faithful Aldhelm;
Bid him endure –
SIGEBERT. He's here! –

Scene xxiii

EDWY. ORGAR. SIGEBERT. ERNULF. ELTRUDA. PEASANT. SOLDIERS.

Enter ALDHELM.

ELTRUDA. Ah! all too late! –
O Aldhelm, thou hast hasten'd to a scene
To tear thy Heart, to rend thy fortitude,
And level them with the poor feeble Eltrud's!
ALDHELM. Ye heavenly Powers! – My King! –
EDWY. Aldhelm? – is't thee? –
ALDHELM. Alas! – for this sped I my perilous march? –
EDWY. Guide of my Youth! my Friend! my Counsellor!
Come close, thou holy man! O bend thee down,
And with thy last kind blessing sooth my exit! –
ALDHELM. My royal master! – the poor Aldhelm's blessing –
Alas! – my voice is choak'd! –
EDWY. Look there! look there!
See who lies there – and grieve for Me!
See thy lost Queen – the mistress of my Heart, –

xxii: an insertion, replacing 'xix'.
V xxii 3 *faithful*: an insertion, replacing 'noble'.
V xxii 3 *Aldhelm*: 'Commend me, too; tell him I die resign'd' is deleted.
xxiii: an insertion, replacing 'xx'.
V xxiii 3 *thy*: 'firm knit' is deleted.
V xxiii 12 *Me*: 'no more' is deleted.

The Partner of it's woes – O Aldhelm, see!
Murder'd – distain'd with blood! –
ALDHELM. Dire, dreadful Sight! –
EDWY. Cast there thy Eyes – on that celestial visage, –
Tell me she is blest! tell me her gentle soul –
O Aldhelm! – Friend rever'd! Benign Preceptor! –
Farewell! – I come! – my angel Elgiva! –

Dies.

ELTRUDA. He's Dead! he's Dead! –
ALDHELM. So soon? – so early flown? –
And is it mine, O King, to see Thee go?
Alas, I thought not this! – My last kind blessing?
My virtuous Pupil! my unblemish'd Prince!
Yes, Aldhelm's blessing is thy well-earnt mede. –
Lives there the Infidel can view this scene
Of slaughter'd innocence, – these mangled bodies, –
See Youth untimely fell'd, Virtue oppress'd,
And Guilt victorious, yet retain a doubt
That Retribution waits for all above?
There shall we see Virtue transparent shine,
All Right refulgent, and all Wrong inverted;
All purity upheld; all crimes remov'd
Ev'n from the power to give the pain of pity: –
Edwy and Elgiva shall there arise
To Thrones of permanence, and Crowns immortal!

V xxiii 24 *mede*: four illegible lines are deleted.

EPILOGUE

The dismal ditty o'er, and Tragic din
Of Sighs, groans, Swords and Screams, awhile shut in
Permit some lighter that before we part.
  I come not here with soft insidious art
To ask your fair protection for the Play;
Ere I appear, its doom has found its way
In piercing hisses, or in generous claps,
Through shrieking Catcalls – or by friendly raps.
I'm therefore here not yet to praise, nor scout it;
But 'talk about it, Goddess, and about it.'
  What shall we say to Edwy and his Rib?
The Boy was sure a Driv'ler – a mere bib,
To make a rout that cost him Crown and life,
To keep a Horse? – a mistress? – no, a Wife!
  His taste was all his own, we must allow;
How many may there be I see just now
For the same end would gallop the same road?
  Not a Hand moves! so did my mind forbode. –
Forgive me Ladies! men are such sad Rogues.
And truth, sometimes, peeps forth in Epilogues.
How long had they been married? cries some wag:
Was all the Honey sipt? – the Bride a Hag?
Had her sweet Temper weather'd winter's cold
Yet never warm'd it with one genial Scold?
Hard in her Ear did Tyrant mandates blow
Yet on her Cheek no livelier ruby glow!
Had she been whirl'd from some dear fav'rite whim
Yet met the Tempest like some Zephir trim?
When Passion's Lightning has approach'd her nigh
Has nought electric sparkled from her Eye?
Or when quick rage it's thunderbolt has flung

Epilogue: the text is taken from the Larpent manuscript, the only surviving version. The epilogue was spoken by Sarah Siddons.
1 *dismal ditty*: the reviewer for *The Times* (23 March 1795) terms this a 'truism'.
3 *that*: thus in the ms., although the line makes little sense; a noun seems called for after 'lighter'.
8 *Catcalls*: the unpleasant sound made by squeaking instruments to express disapprobation in the theatres.
10 *'talk about it, Goddess, and about it'*: from Alexander Pope's *The Dunciad*, Book IV (1742, 1743), l. 252:
  And write about it, Goddess, and about it.
28 *Zephir*: a mild, gentle breeze or wind.

Has no responsive rattling – Seiz'd her Tongue?
– Nay, if 'twere so, what wonder will't create
Some risk were run to rescue the tired Mate?
Fie, fie! what wicked Bachelor is there
Whose scornful lip, knit brow, and mocking leer
Announce the Virtuous Fair so high in praise
So sought, so lov'd, – the Wife of a few days!
O what a charming change in those grim vows
Which bind for best and worst each prison'd Spouse.
If, in the bond, the worst might be suppress'd
And either side assert where ends the best!
Men, blameless, then, might gently wile away
Their idle Hours in soft philandering play,
And Ladies – if the successors bespeak
Appear as Brides most virtuous – once a Week.
What had those ancient pontiffs in their mind
When they made Laws to shackle all mankind
In Matrimonial Vows? now with permission
To wed or this or that; now Prohibition
To take or one or t'other; while good sense
And Piety alike receiv'd offence
Since all was arbitrary, changing, vague,
Fram'd, one would think, but to Perplex and plague.
Could you not hang up Dunstans by the Dozen?
Make such a rout for marrying a fair Cousin?
O in these happier days, enlighten'd times,
When superstition fabricates no crimes,
When the Soul's energy, ere cramp'd by bands
Prizes the faith which first it understands,
May the fair Kinswoman of our Royal Race
Far hence the Bigot frown eternal chace:
Shew Reason and religion may unite,
Pursue the Example fair before her Sight,
And blessing Those in whom we all take part
Be blest herself by every British heart.

61 *fair Kinswoman of our Royal Race*: Caroline of Brunswick, who married her cousin George, eldest son of George III and heir to the throne, on 8 April 1795. The couple, however, separated a year later. The reviewer for *The Times* admired the 'well turned compliment to the Royal Bride Elect'.

Hubert De Vere

A Pastoral Tragedy

HUBERT DE VERE

There are two surviving manuscripts of *Hubert De Vere*, both in the Berg Collection. The first, in Burney's hand, consists of 131 pages sewn into notebooks, together with 8 loose scraps. Its original subtitle, 'A Tragedy', was at some time changed to 'A Dramatic Tale, in five parts'. It is very heavily corrected (black ink covering brown), while other revisions are pinned or pasted to the pages on separate pieces of paper. The second manuscript, also in Burney's hand, is subtitled 'A Pastoral Tragedy'. It consists of 110 pages sewn together in five notebooks, one for each act. This copy contains many further corrections, as well as revisions pinned or pasted to the pages. It is, however, much the more legible of the two.

In both copies, Burney has indicated the time taken for a dramatic reading. In the earlier copy, which provides a fuller record, the times are given at the beginning of each Part: 18 minutes for Part I, 30 minutes for II, 35 minutes for III, 30 minutes for IV, and 33 minutes for V. The total of 2 hours, 26 minutes would make the play of a reasonable length.[a]

The second holograph is used as the copy-text for the present edition. Burney's numerous revisions to this copy are recorded in the textual notes, which indicate when the revisions are pasteovers. Most of the obliterated passages have been recovered. Some would be illegible, were it not for the other copy; passages obscured in the later version can often be deciphered through comparison with the corresponding passages in the earlier text. The play has not been previously published.

Burney began writing *Hubert De Vere* in 1790, during her last year of court service. She had scarcely completed the first draft of *Edwy and Elgiva* in mid-August when 'imagination seized upon another subject for another tragedy' (*Diary and Letters*, IV, 413). By 5 June 1791 she seems to have written the first draft of *Hubert De Vere*, since in a diary entry of that month she contrasts her third tragedy, *The Siege of Pevensey*, 'only generally sketched as to plan and character', with her first and second, *Edwy and Elgiva* and *Hubert De Vere*, which were in a more advanced state of completion (*Diary and Letters*, IV, 478).

On 3 July 1793, a notice in the daily newspaper the *Oracle* stated: 'Miss Burney's pen will retrieve the Stage, degraded beyond bearing by the tricking trash which our Harlequin Writers have forced upon the Public'. Two days later, the same paper announced that 'Kemble accepted Miss

[a] Joyce Hemlow notes that Burney read and timed both *Hubert De Vere* and *The Siege of Pevensey* around 1836, when she was 84 or 85 ('Fanny Burney: Playwright', p. 180).

Burney's Tragedy in the proper way – The possession of such acknowledged talents is superior to all private opinion either way. The House is open to her of *course*'. These remarks attracted the attention of Burney's father, who wrote to his son Charles in a letter of 16 July 1793: 'Pray ... what was meant by the 2 Paragraphs in the Oracle abt Fanny's play? ... The play wch you returned after Kemble had seen it is in my possession, & I have never heard what was his *Vardy thereon*'.[a]

That the '*Vardy*' or verdict of John Philip Kemble, manager of the Drury Lane theatre, was positive and that the play in question is *Hubert De Vere* is shown by a much later letter from Burney to her father of 26 January 1797. Here she recollects that Kemble had once accepted *Hubert De Vere* for production at Drury Lane, but that she herself had decided to withdraw the play 'before he had time to shew it to Sheridan, from preferring to make trial of *Edwy and Elgiva* because it was more dramatic' (*Journals and Letters*, III, 258). Dr Burney, she reminds him, had read *Hubert De Vere* at that time, '& it lost you, you told me, a night's rest'. On rereading the 'melancholy ditty' she has decided 'never to essay it for representation'. Instead she will 'restore it to its first form, that of a Tale in Dialogue, & only revive, & endeavour to make it readable for a fire-side'. In the same letter, however, Burney writes that the 'abominable second' of the 'abominable Mr. Lewis makes me waver from my original intent of *lying fallow* for some years to come. And therefore I shall prepare my Dramatic Tale as well as I can, & then ponder & consult upon its production probabilities'. Although the allusion to Lewis (possibly Matthew Lewis, notorious for his novel *The Monk* (1796)) is unclear, Burney seems to be reconsidering the possibility of having *Hubert De Vere* produced on stage.

In his reply of 6 February 1797, Dr Burney wrote that he was 'glad you have taken up your *Tale in Dialogue* – It pleased me, I remember; but seemed too simple for our stage – but as a Tale, I have no doubt but you will make [it] most pleasantly interesting. On! On!'.[b] Burney seems to have accepted his advice, abandoning her thoughts of a stage production but continuing to revise the text for publication as a closet drama. In a letter of 26 February, she told her father that she was still revising the 'Tale in Dialogue', and offered to send a copy to his close friend Frances Crewe 'before publication' (*Journals and Letters*, III, 280).

Burney continued to revise the two versions of *Hubert De Vere* for some forty-five years. The 'Dramatic Tale' manuscript has corrections on a letter fragment postmarked 1818; the 'Pastoral Tragedy' version has corrections on letters bearing postmarks of 1824, 1830, and 1835. Burney was clearly working on both manuscripts late in her life, but completed her work on neither.

There are no references to *Hubert De Vere* in Burney's letters or

[a] Unpublished letter, ms. owned by H.L. Platnauer; transcription from the Burney Papers, McGill University.

[b] Unpublished letter; Berg Collection.

journals after 1797, and despite her wish to see it in print she seems never to have shown it to a publisher. In a scrap written when she was timing the play, probably in 1836, she expresses positive sentiments about Act V:

> Altogether excellence greatly rising wants chiefly poeticality but very touching very elevating very saddening & worth rectifying De Mowbray too much of crime unless earlier marked to repentance & remorse & only called off for safety. make that change & all his soliloquys will be extra good & new Devere admirable Geralda exemplory Cerulia a little modified & her disapperance more & marked Great interest awakened.[a]

[a] 'Miscellaneous Pieces of Manuscript, 1772–1828', folder IV, Berg Collection.

PERSONS OF THE DRAMA

THE BARON DE MOWBRAY
HUBERT DE VERE
BEAUCHAMP, GIFFORD } Vassals of the Lord De Mowbray
GEOFFRY, a Shepherd
HUMPHRY, an Husbandman
MESSENGER

GERALDA, Niece of the Lord De Mowbray
CERULIA, AGATHA, PHEBE, AMY } Shepherdesses
VILLAGE MAIDENS

The Scene passes in the Isle of Wight,
In the Reign of King John.

ACT I

The Landscape represents a Heath,
with a view of the Sea on the back Ground.

HUBERT DE VERE *appears at a distance, his arms folded, his Eyes fixed on the Earth. As he passes slowly out of sight,* GIFFORD *and* BEAUCHAMP *advance.*

BEAUCHAMP. O! I can act this treacherous part no more,
So noble is the man, so dire his fate!
I came to watch, or fabricate, some crime; –
But Enmity itself must yield to Misery
Pursu'd by Vengeance, though by Guilt unsullied.
Triumphant Virtue! what shall cope with thee?
Oppression makes thee firmer; multiplies
Thy means, resources, energies. Thy bulwarks
Nor Art can sap, nor slander undermine;
Strong in their hold, they brave even Passion's storm,
And turn Malignity to Admiration.
GIFFORD. 'Tis not for us, the Vassals of De Mowbray,
To judge his mandates; or to weigh their rights:
He hates De Vere; he wills to ruin him:
Our charge is to precipitate his fall:
If crime there be, it hangs on who commands.
BEAUCHAMP. How savage is that hate which, unsubdued,
Views biting Wrong, sharp Grief, and fell Disgrace!
Hubert De Vere, the first on Valour's list,
Exiled, unheard, by John, to this small Isle;
Betray'd by her he lov'd, the false Geralda;
And wasting here his life, his peace, his fame –
Is yet the object of De Mowbray's rage,
Though dark it's purpose as it's cause mysterious.
GIFFORD. Credence disclaims the soaring Soul of Hubert
Should thus be sunk for a frail, futile Woman.
Geralda, wedded to the potent Glanville,
The Lord of Wealth, as Hubert is of Honour,

I 20 *John*: son of Henry II and Eleanor of Aquitaine; reigned 1167–1216.

ACT I: 'Scene I' is deleted.
I 1 s.d. *with a view ... back Ground*: an insertion, replacing an illegible line.

Merits no more regret, but haughtiest scorn.
BEAUCHAMP. Scorn? What is Pride where Love usurps the Empire?
Food for it's Mockery, it's scoffing gambols,
It's playful wiles, and amorous defiance.
Love makes a Jubilee of Pride's high bearing
Till, tam'd and humbled by the wanton tyrant,
It's lofty height is cast at Beauty's feet.
Geralda, too, De Mowbray's Niece – perchance
Some foul constraint has work'd this seeming change.
GIFFORD. Portentous of some evil is the silence
And fearful secresy De Mowbray guards.
'Tis now two years since here we have been station'd –
BEAUCHAMP. And say! what have we found oblique in Hubert?
Strait is his course, inflexibly direct
As his high thoughts; though, lost to all mankind,
He brood in deadly wrath, o'er wrongs and woes –
Unheard of Wrongs! unutterable Woes!
GIFFORD. Some latent project stimulates De Mowbray;
Why on Cerulia fix to lure De Vere
Again to Passion's slavery? Has De Mowbray
Beheld that village Maiden? – Why Cerulia?
BEAUCHAMP. Not there the marvel lies, for as she's lovliest
She soonest might eclipse the fair Geralda;
But most he points the rancour of a foe
That he would taint the high born Hubert's lineage
By union with a maid unknown, unportion'd,
An humble rustic Orphan.
GIFFORD. Union? True;
De Mowbray still enjoins we guard her honour.
BEAUCHAMP. Were Hubert, by foul arts, to blight her fame
Pity from him would turn to rest on her:
Hubert De Vere is form'd of finer mould;
Though work'd to phrenzy, alter'd, chang'd, morose,
Shunning society, detesting light;
Still is his fury but of outward guise;
Still blameless is his life, severe in virtue;

I 31 *Food*: before 'Food', 'More' is deleted.
I 31 *it's*: an insertion, replacing 'whose'.
I 32 *defiance*: 'Laugh at it's ire, sport with it's arrogance' is deleted.
I 33 *Love*: an insertion, replacing 'Pride'.
I 33 *Pride's*: an insertion, replacing 'it's'.
I 34 *wanton tyrant*: an insertion, replacing 'sorrow's tyranny'.
I 46 *Some latent ... De Mowbray*: an insertion, replacing 'What latent view lurks in De Mowbray's breast?'.
I 48–49 *Has De Mowbray ... Why Cerulia*: an insertion, replacing 'Why Cerulia? O [two illegible words] where has De Mowbray seen her?'.

Still, though of aspect fierce, his heart is tender!
GIFFORD. Scarce fifteen years had young Cerulia told
When to this Isle mysterious banishment,
In silent gloom, condemn'd De Vere; since when
How have we urg'd with Hope, with Flattery foster'd
Her fond, unwearied chace of the stern Exile!
BEAUCHAMP. But, faithful to the rage of injur'd love,
Though Falsehood has dissolv'd Hope's fairy fabric,
He sees, or heeds not, this fair paragon,
Untouch'd by tenderness, to Beauty callous.
Who but would wonder that beheld her charms,
And saw with what young, unresisting softness
The artless Virgin gives him all her soul.
GIFFORD. Yes, she surpasses even De Mowbray's views;
He meant her to ensnare the heart of Hubert;
But, lighter to allure, her own's enslav'd.
BEAUCHAMP. This way he bends his ruminating walk:
Let us retire.
GIFFORD. Be't mine, once more, to obey
De Mowbray's charge, and urge the fair Cerulia
Hither, with winning Love, to speed her steps.

Exeunt.

Enter HUBERT DE VERE.

DE VERE. Aught than the lure of wealth were less ignoble –
For wealth – oh sordid perjury! – break her vows! –
Throw her fair Form into another's arms –
Inhuman infamy! – Glanville, beware!
Fair as it seems, 'tis hollow; all within
Of soul, or mental faculty, is wanting;
All nobler splendour of essential worth,
Founded on Honour, Faith, and God-like Truth.
Fair attributes, that once I thought Geralda's,
Say, whither, are ye fled? – Let me forget her! –
Or rather let my thoughts hang o'er her image

I 64 *Still . . . heart is tender*: an insertion, replacing 'Still kind in every deed, though fierce his aspect'.
I 78 *ensnare*: an insertion, replacing 'enslave'.
I 82 *fair Cerulia*: an insertion, replacing 'tender maid'.
I 87 *beware*: an insertion, replacing two illegible words.
I 90 *worth*: 'All basis of magnificence of mind' is deleted.
I 92 *Fair*: an insertion, replacing 'O'.
I 92 *Geralda's*: two lines are deleted:
How are ye fall'n! sunk in the low abyss
Of Avarice and Vanity, – fell miners.
I 93 *Say . . . fled*: an insertion, replacing 'Of all that's Right and High'.
I 94 *Or*: an insertion, replacing 'Not'.

Till every feature picture some offence
And shew her beauty hideous. – Oh those Eyes! –
Lustrous with sentiment, with Honour beaming –
With Honour? – barbarous hypocrite! – Black! – Black!
Yet blacker her duplicity than in Love;
To quit me – though life parted thence from joy,
And every faculty of hated being
Shrunk from its post at the perfidious tidings, –
But heralded her plot; – even to wed Glanville –
A deed to whiten every crime yet known! –
Was still availess to complete my fall;
Her ravenous Hate demands my Fame! – Detraction
Hiding its hideous Form in Mystery,
Flying Enquiry, while assailing Truth,
Proscribes me from my King, my House, my Honours,
And gives the stamp of Guilt, where every Daemon
Of Earth – or Hell – might search in vain the proof.
Ungrateful John! – so lightly to discard me!
Oh harder less to brook is every misery
On which the Sun can set – Labour, – pain – Ruin –
Than one slight tinge of Shame, though it's faint hue
But by surmize be colour'd. –

CERULIA (*from within*). Ah! – there he is!

Enter CERULIA, *gently. She stands aloof.*

How noble is his mien! – but oh how awful!

DE VERE (*perceiving her*). Why will this lovely Innocent pursue me?

He folds his arms, and walks away.

CERULIA (*coming forward*). Alas, he is gone! Oh cruel, faithless Hope,
O'er-shadow'd ever more with Disappointment!
Why, Gifford, why awake the soft delusion
Whose dear expectance mantles to my Heart?

I 96 *Eyes*: three lines are deleted:
Oh! they were falser than her breast! They look'd
As if stern scorn had petrified the miscreant
Who durst pollute the air she breath'd with treachery.

I 98 *Black! – Black*: an insertion, replacing 'O there'.

I 100 *from*: an insertion, replacing 'with'.

I 106 *Her ravenous ... Detraction*: an insertion, replacing 'My fame must [an illegible word] her [two illegible words] hate – Detraction'.

I 107 *it's hidious*: an insertion, replacing two illegible words.

I 119 *Alas, he is ... Hope*: an insertion, replacing 'Ah, he is gone! Art thou o radiant Hope'.

I 120 *with*: an insertion, replacing 'by'.

I 121 *awake the soft*: an insertion, replacing 'urge on the dear'.

Vain Heart! Vain Hope! back to your lowly cells! –
What though the Red-breast may the Sparrow charm,
No Red-breast will the Sparrow's chirping heed!
What though the Dove may fascinate the Swallow,
No Dove will hail the Swallow's humble greeting!

Enter AGATHA.

AGATHA. Cerulia! but now I met De Vere
Wearing upon his brow a scowl, to teach
Fear ev'n to valour.
CERULIA. All the Village fear him,
Save poor Cerulia; yet none know why:
Can goodness raise alarm? Can generous pity,
Or heavenly Charity, excite mistrust?
Yet who is good, or bountiful, or pitying
If Hubert be not he?
AGATHA. True; but his pride –
CERULIA. Nay, nay, – dear Agatha, – what thou deem'st pride
Seems to me Nobleness and high desert;
A grander, loftier, – not a prouder Nature.
AGATHA. Why lives he still so distant, and so lone? –
CERULIA. Thou dost not mark distress from haughtiness.
For solitude he shuns us; not from scorn.
Aloof, I have long observed him, and still seen
Something the more to charm – and yet to awe. –
Through that stern air, O what a Soul of pity!
Through all those rugged ways, what grace superiour!
AGATHA. Gifford still holds that, caught by thy soft Eyes,
Love, lordlier even than Hubert –
CERULIA. Ah! not so! –
Hubert De Vere can ne'er love poor Cerulia:
Yet to his cot how oft has Gifford urg'd me

I 124 *Red-breast*: robin.
I 149 *cot*: cottage.

I 122–23 *Whose dear ... lowly cells*: an insertion, replacing:
That mantles to my heart with false Felicity
And glowing thoughts, that breathe Hubert De Vere.
I 125 *will*: an insertion, replacing 'must'.
I 127 *will*: an insertion, replacing 'would'.
I 116 s.d.-130 *Enter* CERULIA, *gently ... Village fear him*: a pasteover.
I 131 *Save poor ... why*: an insertion, replacing 'Save me; yet why I know not – nor do they'.
I 141 *scorn*: an insertion, replacing 'hatred'.
I 142 *have*: an illegible word is deleted.
I 142 *Aloof ... seen*: an insertion, replacing two variant lines: 'Day after Day I have watch'd him, and still seen' and 'Observing still I have watch'd him, and still found –'.

With my late foster-Father, good old Eustace?
Thither he'd prompt us on, with fresh persuasion
Now for ourselves to plead; now, for some neighbour:
De Vere was still the same, still good and bounteous;
But yet – ne'er ask'd our stay! – ne'er bid us welcome! –
Ne'er spoke of our return! – though oft, at parting,
Be she thy Morning Star, old man! he'd cry,
Hail her soft beams, and watch no cloud insidious
Obscure her from thy sight, or dim her lustre;
And Thou, young Maid, brighten his waning years
With rays of virgin Kindness, pure, etherial!

AGATHA. Thy timid Nature Gifford calls thy bane;
And bids me quell thy doubts, and raise thy hopes.
De Vere, he cries, adores thee; but, being sad,
He needs much gentle soothing and soft seeming.

CERULIA. Ah, could I humbly, and at distance serve him
Agatha, Oh what joy! – to see him daily;
Sometimes to hear his voice; – sometimes, perchance,
To meet his beaming Eye! – Oh Agatha! –
To witness all his deeds of heavenly bounty;
To be, perhaps, his Almoner, distribute
From his own noble hand, his kind donations,
And, through his blessings, become bless'd myself! –

Enter GIFFORD.

GIFFORD. Haste, gentle Maid, pursue thy prosperous fortune;
Towards the tall Cliff the noble Hubert wanders;
Speed thou thy way, and glad him with thy sight.

CERULIA. No Gifford, no! – I follow him no more!
Alas, he heeds me not!

GIFFORD. 'Tis Grief's perverseness.
He aims to shun thee, yet he pines without thee.
When thou art present, 'tis his exile pains him;
When thou art absent, every Sigh is thine.

CERULIA. Mine? – I would die to give him peace and Joy! –
Towards the cliff, said'st thou? – Should it anger him! –
What if I take the Basket I have woven
And bear him Flowers and Fruit?

GIFFORD. Nothing so welcome.
Come, gentle maid, I'll guide thee to his presence.

Exit GIFFORD *with* CERULIA.

I 153 *bounteous*: an insertion, replacing 'generous'.
I 175 *glad*: an insertion, replacing 'sooth'.
I 178 *shun thee*: an insertion, replacing 'while'.

AGATHA. So gloomy he, and She so meekly sweet –
Is't not some Sorcerer charms away her heart? –
Blinding, meanwhile, De Vere, that, though he gaze,
He keep, – I marvel how! – his own untouch'd?
If as the Beech he is tall, erect, and strong,
She as the Hawthorn blooms with the fresh fragrance
That greets the opening Spring, and spreads it's perfume.
'Twere manly to protect the weaker plant
Whose balmy odour soon his griefs would chace, –
Or turn to joy.

Enter GEOFFRY.

My Geoffry dear! –
GEOFFRY. My Agatha,
Has not an unknown lady pass'd this way,
Lighter of motion than the timorous Hare,
Lovelier in beauty than the turtle Dove?
AGATHA. Whom may this be, my Geoffry?
GEOFFRY. A fair Stranger.
Crossing, but now, yon rugged path, that winds
Down to the sandy beach, a fearful scream
Sudden mine Ear affrighted. Swift I ran
Pursuing it's shrill sound, which guided me
To where an helpless lady sought release
From three stout cowards, who, with dastard strength
That knew her native weakness, dragg'd her on
Heartless to cries or pray'rs. –
AGATHA. My Geoffry dear!
Have men two natures? These are wide from Thee
As seas from skies, that never can commix.
How didst thou chastise them? how free the lady?
GEOFFRY. To combat with them singly on I rush'd;
But they were arm'd, and dar'd me: back I hied
To claim the vigorous force of sinewy Humphry;
While two brave Youths, scorning to lag behind,
Quick follow'd to the chace.
AGATHA. 'Twas so decreed;

I 187 *charms*: an insertion, replacing 'wiles'.
I 194–95 *chace ... joy*: an insertion, replacing four illegible words.
I 197 *Lighter*: an insertion, replacing 'Swifter'.
I 197 *timorous*: an insertion, replacing 'frighted'.
I 199 *Whom may ... Geoffry*: an insertion, replacing 'Of whom thus speaks't thou, Geoffry?'.
I 205 *three*: an illegible word is deleted.
I 212 *dar'd*: an insertion, replacing 'brav'd'.
I 215 *decreed*: an insertion, replacing 'allotted'.

Oh high in Fortune's favour ever stands
The Damsel's champion; – Friend to the helpless!

GEOFFRY. Like ravenous Wolves for rage we sprang at them
With our stout weapons of good husbandry,
And foil'd their arms, and drove them from the field.

AGATHA. My Geoffry brave! – But say, how fares the lady?

GIFFORD. Why there's the ill, for as we fought, she fled.
Whither we know not. On her search we parted,
Each taking different routes.

AGATHA. Alas, poor lady!
O, I will look the Island o'er and o'er;
Cerulia, too, shall look; for well she knows
Each mossy Glade, each Wood, each tow'ring Hill,
Each lonely valley, and each secret path
That Fear might seek for shelter. Poor Cerulia
Has learnt them all in following proud De Vere.

Enter HUMPHRY.

Welcome, brave Humphry!

GEOFFRY. How has sped thy chace?

HUMPHRY. E'en as we ne'er had budg'd. Deep in the Wood
The lady is fled. I met but now a Youth
Who, fishing on the strand, beheld the vessel
In which she gain'd the Isle. He has unfolded
That scarce she landed ere these base poltroons
Assail'd, and forced her off; while their compeers
Captur'd her Yatcht, imprisoned her attendants,
And sail'd away with them to distant Shores.
Another vessel near the coast is mark'd

I 236 *poltroons*: cowards, wretches.
I 238 *Yatcht*: variant of yacht.

I 216–17 *Oh high ... helpless*: an insertion, replacing:
For high – oh high in Fortune's fairest favour,
Kind Fortune ne'er forsakes the damsel's Friend
Friend to the helpless stands the Damsel's champion.
I 221 *But say ... lady*: an insertion, replacing:
I knew all luck was thine,
I knew upon thy birth the planets smil'd.
Thou coulds't not fail to win. How fares the lady?
I 227 *Each mossy ... Hill*: an insertion, replacing 'And better far than I, each glade, each wood'.
I 231 *chace*: an illegible line is written over.
I 236 *base poltroons*: an insertion, replacing 'puny traitors'.
I 238–39 *Captur'd ... Shores*: an insertion, replacing:
Fell'd her attendants; bound them; seiz'd her Yacht,
And sail'd away with it to some distant shore.

To which, perchance, they meant to bear the lady.
AGATHA. Soft are the Rocks, smooth is the briny surge,
To Man's ungentle heart, by evil tainted.
O, rather midst the tempest would I sail
On the huge ocean, while each swelling billow
Foams, white and frothy, o'er the crazy vessel,
Threat'ning to sink it in the unknown gulph;
Or dash it on the shoals, a pitiless wreck
For Famine and for Death, than yield my trust
To him who feels not for his neighbour's sufferings.
GEOFFRY. Unkindness is a sin 'gainst Nature, Agatha:
All fellow-creatures born, howe'er we are bred,
Should we not live in gentle Brotherhood?
Each that is sorrowing calling for relief,
Each that is prosperous, willing to bestow?
Even De Vere, who shuns the haunts of men,
Turns, at the voice of woe with kindly succour.
AGATHA. 'Tis then alone I like to look at him.
In offering alms his Eyes lose half their fierceness,
And his dark, angry brow unbends to sorrow.
GEOFFRY. On then, my friends; and from those pouncing Hawkes
Rescue their quivering prey. Our fair success
Shall foster and sustain her; – or our kindness
Back turning on ourselves, sooth our defeat.
AGATHA. Right sure; but ne'er did peace enwrap the bosom,
Nor favouring fortune light upon the roof
Where Charity's soft voice implor'd in vain,
Where fostering sympathy ne'er sooth'd distress,
Where Misery call'd – unheard! – for help and pity.

Exeunt.

I 241 *lady*: an insertion, replacing 'captive'.
I 244 *O ... sail*: an insertion, replacing 'O rather would I venture in a Storm'.
I 245 *huge*: an insertion, replacing 'great'.
I 260 *sorrow*: an insertion, replacing 'softness'.
I 269 *Where Misery ... pity*: an insertion, replacing 'Where misery call'd in vain for help and pity'.

ACT II

Rural Scenery, with distant Hills.

BEAUCHAMP *and* GIFFORD.

GIFFORD. The probe of Disappointment ulcerates,
Not purifies, the wounds of lost Ambition: –
In gloom vindictive sinks the proud De Vere.
The maid, with all her beauty, all her fondness,
Expiring at his feet, might lie unheeded.
BEAUCHAMP. Cease, then, to persecute his constant mind.
Unpitied victim of Despair, Disgust, –
Why rob him of the lone, yet lordly solace
That elevates affliction, conscious firmness?
Let us away, and leave him to his fate.
GIFFORD. How! when the recompense of vigilance
Advancing, meets our grasp? Would'st thou we toil
As Hounds pursue the Hare, which caught, they taste not?
Shall years of Service –

Enter DE MOWBRAY: *suddenly, but cautiously.*

DE MOWBRAY (*in a low voice*). Ha! – Beauchamp? – Gifford? –
GIFFORD. The Lord De Mowbray!
DE MOWBRAY. Silence! – name me not!
BEAUCHAMP. What hither brings my lord? And why disguised?
DE MOWBRAY. Motives of deepest weight enforce my presence.
Glanville is dead.
BEAUCHAMP. Lady Geralda, then,
Again is free?
DE MOWBRAY. Free? – Ay; – What would'st thou say?
Was she not free when she would wed with Glanville? –
Would'st thou infer –
BEAUCHAMP. My lord, I crave your pardon!

II 1–2 *The probe ... Ambition*: an insertion, replacing:
No adamant so hard, no rock so rough
As disappointment's probe to high Ambition.
II 7 *Unpitied ... Disgust*: an insertion, replacing 'Victim already of Disgust, Despair'.
II 9 *affliction*: an insertion, replacing 'Afflicts, and'.
II 13 *taste not*: an insertion, replacing 'touch'.
II 14 *Shall*: an insertion, replacing 'Two long'.
II 14 s.d. *DE MOWBRAY*: 'rather' is deleted.

De Mowbray. Well, – let it rest. This niece of mine is fled;
Hither she bent her way.
Beauchamp. And why? – what purpose –
De Mowbray. Perchance she seeks De Vere; but – mark me Well!
They must not meet. Matters of high import
Demand their separation. – Why so curious?
Beauchamp. My lord, I spoke not. –
De Mowbray. Think'st thou Question waits
The sluggish Tongue to note its froward yearnings?
The Eye – each feature is at work, to purloin
Fresh Game – tid bits, for it's voracious cravings.
The crafty Tongue is mistress of its longings,
And knows to vest them in ambiguous phrase;
But in an instant an unruly glance
Lets loose Enquiry – Doubt – Surmize – Distrust – .
What is't thou would'st conjecture?
Beauchamp. Good my lord –
De Mowbray. O, true, – thou didst not speak! – I had forgot; –
Of this no more; vain, idle reasoning, –
Captious and futile. – Change to manlier action.
Haste – fly – breathe not till she be captur'd, – Hence! –
Gifford. My lord, I am gone. –
Beauchamp. Captur'd? – What? – Who, my lord?
De Mowbray. Who? – Glanville's Widow; – said I not? – Geralda.
Beauchamp. If she be landed –
De Mowbray. If? – heard you me not?
Who am I, Sirs – I have had her waylaid, seiz'd;
But she escap'd. Bribery, or some foul art
Enlarg'd her. Without friend, or aid, or guide,
She roves the Woods; but you, to whom this Isle
Must now be Home, she vainly would elude.
Chace her then, quick: – a Vessel skirts the Shore
To speed her hence. – But first, secure her papers! –
Her Letters – Writings – Whatsoe'er ye find;
And, on your fealty, guard her from De Vere! –
But hold! – What further progress makes Cerulia?
Does she attack him?
Beauchamp. O, She needs no urging:
Her heart, unpractic'd, is all tun'd to love.
Caught by his sorrows in the snares of pity,

II 28 *froward*: perverse, refractory.

II 30 *voracious*: an insertion, replacing 'rapacious'.
II 40 *What? – Who*: an insertion, replacing 'Who means'.
II 46 *Woods*: an insertion, replacing 'Isle'.
II 46 *Isle*: an insertion, replacing 'Coast'.
II 47 *Must ... Home*: an insertion, replacing 'Through-out is known'.

A tenderer sympathy involves compassion.
Her artless nature –
DE MOWBRAY (*eagerly*). Have you still preserv'd her
Strict to the path of honour?
BEAUCHAMP. Good my lord,
De Vere is no deluder; and the maiden,
Though not a Rose so attractive or so fair,
Blooms, like the Violet, only in the Shade.
A breeze would ruffle her; a gaze, confound;
And Hubert by avoidance wins that favour
Which, were she sought, her bashful fears would check.
DE MOWBRAY. She loves him, then?
BEAUCHAMP. Oh, more than life or light!
How oft, when we have wrought on aged Eustace
With humble suit to urge her to De Vere,
Have I beheld glow crimson on her cheek
The blushing hue of juvenile delight!
While Hubert, mute, pre-occupied, heart-worn,
Saw but the simple Peasant's messenger,
Gave what was ask'd, yet scarcely knew to whom.
DE MOWBRAY. If you had brought to bear that wish'd alliance
My cares had ceas'd –
BEAUCHAMP. My lord?
DE MOWBRAY. Silence! – And, mark me! –
Their union is my life!
BEAUCHAMP. How would it joy her!
Though, all unconscious of each subtler aim
The timid Maid aspires but to adore him.
DE MOWBRAY. Hence, Beauchamp, scour the Coast. Gifford shall after.
Remember! if Geralda join De Vere
Destruction lights not on my head alone!
Exit BEAUCHAMP.

DE MOWBRAY, GIFFORD.

DE MOWBRAY. Gifford, I deem thee honest – secret – firm; –
But – Beauchamp – may I trust him? – Is he steadfast?
GIFFORD. My lord, he holds not with respect like mine
A Vassel's duty; nor, like me, is prompt
Or active to perform mysterious service;
But still, – I know not that he wants for honour.
DE MOWBRAY. He has ever been too prone to reasoning;
There lies the evil. Will he seek Geralda?
GIFFORD. My lord, no doubt; – but yet –
DE MOWBRAY. Will he give ear
To idle pleadings? – Ha! – Will Tears – and prayers –
Those common wiles of female subtlety –

Unman him? – speak! – be bold! – Fear not! – Denounce him! –
GIFFORD. Beauchamp, my lord, could sooner bear rank insult,
Ev'n from an Equal's arm, than turn aside
From one soft pearly drop on Beautiful cheek.
DE MOWBRAY. Call him then back! – Bid him give o'er the chace! –
Tell him I wish her lost! – abandon'd – free! –
Away without reply! –

Exit GIFFORD.

DE MOWBRAY *alone*.

Avenging Powers!
How Gulph on Gulph tremendous gapes before me,
Watching the first weak, devious step, to hurl me –
Where? – How? – Avaunt! – quit quit my troubled brain,
Ye hideous fancies of eternal doom!
Shall Superstition's despicable arts
Appal the breast that mocks them? Can we feel
While we deride? and tremble where we scorn?
Oh cozenage base of early prejudice!
Adhesion baneful of first fears and habits! –
I dare not crush the dust that Hubert treads –
I fear 'twill turn to particles of fire
To blaze around me, and expose my shame,
Ere, with a kindlier furnace, it consume me!
Here, too, I breathe the air Cerulia breathes –
Oh hapless Orphan! – Not, – oh not to Thee
Be I its poisoner! Not to Thee extend
The infection of my curses – or my deeds!
Oh pure as Aether be'st Thou from contagion
Of the foul atmosphere by which I am parch'd –
Burnt! – inwardly consumed! by which all Nature,
Discolour'd by my more than jaundic'd Eyes
Seems dense vacuity! – impervious gloom!
The light of Heaven obscur'd – or only glaring
With sudden sparks, bursting, red hot and furious,
With shock electric on my shatter'd brain.

Enter GIFFORD.

II 106 *cozenage*: deception, fraud.

II 92 *be bold ... Denounce him*: an insertion, replacing 'demur not. Say the worst'.
II 97 *lost*: an insertion, replacing 'miss'd'.
II 97 *free*: an insertion, replacing 'lost'.
II 100 *first weak ... hurl me*: an insertion, replacing 'least small step awry, to sink me'.
II 113 *Oh*: an insertion, replacing 'Poor'.

Well? – What of Beauchamp?
GIFFORD. Good my lord, I needed
Nor Giant's stride, nor the Doe's bounding flight
To o'ertake him. Where the mind is adverse
The sinews seem unbrac'd; the motion torpid;
And muscular force, that hangs upon volition,
Deserts its post to morbid feebleness.
DE MOWBRAY. On thee, alone, then, let the charge devolve.
What racks had I been spar'd, had prompter zeal
Wedded this Hubert to the peasant Girl! –
Geralda ne'er had sought this loathed spot –
GIFFORD. Her early passion, then, –
DE MOWBRAY. Prithee no comments.
Haste thee away.
GIFFORD. See, where Cerulia rambles!

CERULIA *appears at a distance.*

DE MOWBRAY. Cerulia? – Oh quick shew me! – point her out! –
Cerulia, said'st thou? – Where? – Oh where! –
GIFFORD. My lord,
There! – towards the summit of yon craggy hill.
Do you not see a Form?
DE MOWBRAY. A Form most lovely! –
Great Heaven! – Is that Cerulia?
GIFFORD. You seem mov'd?
Are you not well? – The colour flies your Face.
DE MOWBRAY. Thou art sure? –
GIFFORD. Most sure, my lord.
DE MOWBRAY. Cerulia! –
Unhappy Girl! – And looks she so serene?
GIFFORD. Why not, my lord? Where may the wonder lie?
DE MOWBRAY. Nay, no where – nothing – What a grace surrounds her!
How light, how beautiful, how swift her motions!
Already gone? – Was it a dream? – Speak, Gifford!
Are all my senses right? – Cerulia, said'st thou?
Was that bright, airy Figure young Cerulia?
GIFFORD. The same, my lord. But why are you thus struck?
DE MOWBRAY. I know not. – 'twas Surprize; – a something, – nothing
In certain moods will startle us. 'Tis o'er. –
GIFFORD. Yet still you tremble?
DE MOWBRAY. Visions of times past

II 140 *Is that Cerulia*: an insertion, replacing 'And is that Her'.
II 140 *seem*: 'much' is deleted.
II 142 *sure*: 'tis Her?' is deleted.
II 151–53 *DE MOWBRAY. I know not ... still you tremble*: a pasteover.

Impetuous rush on wayward Memory. –
No more! – Fair be her lot! –
GIFFORD. Visions? – Times past? –
DE MOWBRAY. Fever me not with thy pestiferous yells!
Hie on thy search – yet – could'st thou means devise
Nearer to bring yon maiden to my view?
GIFFORD. She does but trace the steps that Hubert treads:
She shall be here o'th'instant.

Going.

DE MOWBRAY. Mark thou nam'st not
To whom she comes. Simply report a Stranger –
A seer – a cunning man – ay, that will do;
And as some rustic quaint Astrologer
I will exhort her to pursue De Vere.

Exit GIFFORD.

(*alone*). The strings of life had nearly burst asunder! –
Dread Powers! – Was it Remorse – or was it – Instinct
That with my heart thus grappled at her Sight?
Away with Thought! – Begone! – senseless Regret! –
Revelling on fruitless woe! – wanton, invidious. –
Who can call back the past? – O who divines
The curse of Guilt, till Innocence is flown!
Why list we never to the warning voice?
Why prize we virtue most our's no more?
Ah, while in early purity and vigour
Embosom'd we possess it, why not Then
Clasp it to our heart's core, and There enshrine it?
Why wait the pungency of sharp distress,
And pang of keen remorse, to teach it's value? –

II 162 *cunning man*: a fortune-teller; see *Love and Fashion*, above, I, 168.

II 155 *lot*: the rest of the line and the next are deleted:
I am glad she lives
Oh! were her hapless Mother, oh!
II 155 *GIFFORD*: two illegible words are deleted.
II 156 *yells*: an insertion, replacing 'questions'.
II 163 *as*: an insertion, replacing 'in'.
II 163 *rustic*: an insertion, replacing 'village'.
II 167 *with*: an insertion, replacing 'at'.
II 165–70 *The strings of life ... O who divines*: a pasteover.
II 173 *no more*: two lines are deleted:
Why does sweet health, – of body or of soul,
Then seem most fair, most when cast away and wasted?
II 175 *it*: an insertion, replacing 'them'.
II 176 *it*: an insertion, replacing 'them'.
II 176 *it*: an insertion, replacing 'them'.
II 178 *it's*: an insertion, replacing 'their'.

Ye deep and awful thoughts! – torment no longer
With wearing, wasting, – puerile retrospection.
I'll wed her to De Vere, that while I blight
Geralda's wrath, by crushing every hope,
I may appease that dread and injur'd shade
On which I fear to call – and bless Cerulia.

Enter GIFFORD, *with* CERULIA, *holding on her arm a Basket of Fruit and Flowers.*

She comes! – 'Tis she! Sink, sink not, coward heart!
GIFFORD. Take courage, gentle Maid. What canst thou fear?
CERULIA. I fear not for myself! –
GIFFORD. Why, then, this tremor?
CERULIA. Should he forbode some harm to noble Hubert! –
GIFFORD. He will instruct thee what thou hast to hope;
What path to take; and all that will befal thee.
CERULIA. But – will he tell me, too, of Hubert's fate? –
Mine soon may be reveal'd. An unknown Orphan,
Supported, barely, by some hidden bounty;
Dear to no Father; by no Mother own'd;
And stranger to all kindred benediction.
DE MOWBRAY. Oh torture!

He makes a sign to GIFFORD *to depart.*

GIFFORD. See! he wills me hence.

Exit GIFFORD.

DE MOWBRAY, CERULIA.

CERULIA. I tremble!
DE MOWBRAY. Young Maid, approach. I know thy destiny.
CERULIA. Is it not sad?
DE MOWBRAY. No, Damsel: 'tis resplendent.
Last Night, in the great volume of the Heavens,
As Midnight toll'd – I read the stars.
CERULIA. The stars?
DE MOWBRAY. And learnt that – ere the Moon be on the wane
Thou wilt become a Bride.
CERULIA. Ye gracious Powers! –
A Bride? – Ah, say it not!
DE MOWBRAY. 'Tis so decreed.
CERULIA. Then am I doom'd to sorrow ever more!

II 185 *She comes! – 'Tis she!*: an insertion, replacing ''Tis she again!'.

For all that love me – are my heart's aversion;
And All I love – shine high, like Stars, above me!
DE MOWBRAY. Thy lot admits no change. Compose thy mind.
Resistance were a crime.
CERULIA. Adieu, then, Peace!
Adieu blyth Joy, gay Hope, fair Liberty! –
Tell me no more! –

going.

DE MOWBRAY. Hear me, young Damsel, hear!
Prepare thee to be courteous, kind, and docile
To him thy planet fates thee for thy husband.
His Name –
CERULIA. I will not learn – lest I abhor it! –
I will away.

going.

DE MOWBRAY. His Name – with the – eighth Letter –
His Christian Name begins
CERULIA. The Eighth?
DE MOWBRAY. Even so.
CERULIA. The Eighth? – Oh mock me not! – the Eighth – (*smiling*) is –
Hubert's! –
DE MOWBRAY. Thou hast divin'd.
CERULIA. Oh Heaven! – Oh gentle Heaven! –
Delude me not – Make me not o'er to phantoms!
DE MOWBRAY. Greet him with smiles, allure with blandishments;
Heed not his roughness.
CERULIA. I will obey thee all! –
Most Kind Astrologer! – But, – is't a charm? –
Am I within a spell? – I must not trust thee:
Thou art some Enchanter, who would try thy art
On a weak Maiden's vain credulity.
DE MOWBRAY. It is ordain'd. Thy lordly Mate is Hubert.
CERULIA. Oh do not make me wild! – Spare, Kind Astrologer!
Such quick transition from the shades of Night
To Day in all its beauty, all its glory! –
Is it no trance? – Am I Cerulia still? –
Cerulia, and yet happy? – Oh soft change!
It mantles to my heart! – Where are my sorrows? –

II 206 *shine ... above me*: an insertion, replacing 'alas! – scorn to remark me'.
II 207–08 *Thy lot ... crime*: an insertion, replacing:
Resistance were but vain. Compose thy mind;
Thy lot admits no change.
II 218 *Delude ... phantoms!*: an insertion, replacing first 'What sayst thou, good Astrologer? – What meanst thou?' and second 'I alive? – make me not o'er to phantoms!'.
II 226 *CERULIA*: an illegible word is deleted.

Hast thou dissolv'd them, Kind Astrologer?
Are they all melted in a flood of bliss?
My Terrours? – all to Courage turn'd? – My Doubts
That never rose to Hopes! where? – where? – My Fears
That dormant kept my very wishes! – Where? –
I see not, feel not, scarce conceive a Grief –
Though, till thy gentle augury, Doubt – and Fear
Seem'd all my Being! Oh, most Kind Astrologer,
To Thee be every Planet mild, benign!
On Thee fall every blessing thou hast shower'd
On me, poor Orphan! – found without a hope,
But left – without a wish!

DE MOWBRAY. My Soul is pierc'd! –
Damsel, farewell. My Planet guides me hence.
Mark my Decree! And claim thy destined Partner.

Exit.

CERULIA. Am I not new-created? – Form'd of air?
Freed from all Earthy weight? – Oh, had I wings!
How would I soar on high! My swelling bosom
Pants for some finer, lighter, purer Region,
Where it might heave with less tumultuous Joy. –
Yet, Hubert here abides! – Oh here abide
Cerulia, then, – for Hubert is the Star
Through whose bright beams my every ill is vanish'd:
Whose radiance has dispers'd the chill, cold mists,
And mental snows that flak'd upon my breast,
To Ice congealing every fond idea
That breath'd a thought I was not born for woe!

Enter AGATHA.

Oh Agatha, why wert not by my side?
Thy kindly sympathy had calm'd my spirits.
Thou know'st not, Agatha, the Fate's decree? –
Thou know'st not what the Astrologer reveal'd? –
Know'st not – this grand, vast, mighty universe
Contains not, in its various bounds, a Being
So happy – as the desolate Cerulia!

AGATHA. The Astrologer? Oh why was I away?

II 238 *Though, till ... Fear*: an insertion, replacing 'Though [three or four illegible words] Doubt, Terror, Fear'.
II 247 *Oh ... wings*: an insertion, replacing 'Would I had Wings'.
II 248 *How ... high*: an insertion, replacing 'I fain, I fain would fly'.
II 249 *lighter*: an insertion, replacing 'fresher'.
II 262 *mighty*: an insertion, replacing 'high'.

Alas, he might have known when Geoffry's Father
Would cease to cross his generous son's affections.
CERULIA. Tell me, dear Agatha, is not the air
Newly embalm'd with odours fresh, celestial?
See'st thou not o'er yon cliffs the azure Sky
Softly transparent with gay liquid lustre?
Feel'st thou not light thy Spirits? unencumber'd
With troublous cares, pains, wants, or vain desires,
Such as till Now commixt our Smiles with Tears?
Dear Agatha, tell me thou shar'st these sweets,
These Balms? – This bliss? Tell me that not alone
Cerulia's bounding heart thus hails Felicity?

Enter GEOFFRY.

AGATHA. My Geoffry, thou look'st pale? – affrighted? – say;
Hast met the Astrologer?
GEOFFRY. No, Agatha;
But I have nearly reach'd his subtlest art,
And learnt its bearings.
AGATHA. Presumptuous Geoffry! Fie!
GEOFFRY. Nay, I have nearly bade this World adieu
And mounted to the Heavens.
AGATHA. The World adieu?
GEOFFRY. Hard by, while seeking for the lady lost,
Sudden two Ruffians caught me by the arm
And claim'd, with curses loud, their lawful prize:

II 271 *liquid lustre*: the phrase is from William Mason's 'Elegy on the Death of Lady Coventry', l. 15. Burney alludes to this line in *Camilla*, p. 569, and *The Wanderer*, pp. 117 and 766.

II 274 *Tears*: three lines are deleted:
Saddening such joyance, thou each ill recruiting?
But now – enhals't thou not unmingled perfumes?
And riv'st then not, unmixt, prospects ethereal?
II 276 *These ... bliss*: an insertion, replacing 'Partak'st these balms'.
II 280–81 *But I ... bearings*: an insertion, replacing:
But I had nearly grown as wise as he;
Or even wiser.
II 282 *nearly ... adieu*: an insertion, replacing 'almost seen another World'.
II 283 *And mounted to the Heavens*: an insertion, replacing 'To seek the stars'.
II 283 *The World adieu?*: an insertion, replacing 'Another World? Oh, fright me not, dear Geoffry!'.
II 286 *And claim'd ... loud*: an insertion, replacing 'And bade me give them back'.
II 286 *prize*: four lines are deleted:
I knew them then again, and call'd them Cowards;
When quick they both assail'd me with such force
As to the Earth forever more had fell'd me
If –

AGATHA. My Geoffry dear! –
GEOFFRY. But, as they vow'd my death
If I releas'd them not the lady sought,
Forth stood De Vere.
CERULIA (*starting from an extatic reverie*). De Vere? – Hubert De Vere?
Did they then fall on him? – speak, Geoffry, speak!
GEOFFRY. Aloud I call'd for help; stoutly he came:
Oh he is brave as good! of warlike courage,
In Danger valiant, as to Misery bounteous.
CERULIA. Oh Geoffry, keep not thus my heart o'th'Rack –
Yet Now, dear Geoffry, Now thou speak'st him right.
GEOFFRY. With many a well-aim'd stroke he parried them,
Till I could rise and help. Off then they flew;
But, dastard like, in flying hurl'd a stone
Direct at Hubert. –
CERULIA. Hubert? – Oh! – I die! –
Where is he? – Where? – My fated Love? – my Hubert? –

Going.

GEOFFRY. Turn! turn! – behold, he comes! –

Enter DE VERE.

CERULIA. Oh Joy of Joys!
Oh, liv'st thou still, Hubert, my destin'd Hubert? –

She flies to meet him but drops at his feet.

DE VERE. What means this? – Geoffry? – Agatha? – support her! –
Be quick – but gentle.
AGATHA. O! ill-starr'd Cerulia!
My mind misgives me thou wilt die for love!
CERULIA (*rising*). Where – where is Hubert? Say, in pity, say!

II 287–89 *AGATHA ... stood De Vere*: an insertion, replacing:
AGATHA. Oh my Geoffry! Oh my constant love!
Oh! – When the time shall come, my dearest husband, –
Say, art thou safe? –
GEOFFRY. Just as they vow'd my death
If I releas'd them not the lady sought,
De Vere appear'd. –
II 300–01 *My fated Love ... behold, he comes!*: an insertion, replacing:
Guide, guide me to my Hubert!
Where – which way – where shall I find my Hubert?
GEOFFRY. Behold! see where he comes!
II 302 s.d. *drops*: an insertion, replacing 'falls'.
II 303 *support*: an insertion, replacing 'assist'.
II 306 *Say, in pity, say*: an insertion, replacing 'Tell, in pity'.

DE VERE. Calls she on me?
CERULIA (*rapturously*). Sweet Heav'n! 'tis He! – 'tis Hubert! –
My Heart's Elect! – Art thou unhurt? – unwounded? –
DE VERE. I stand amaz'd!
CERULIA. Oh Agatha – oh hide me! –
I sink with shame! –
AGATHA. No, raise thy lovely head!
CERULIA. Oh, never can I look on Hubert more! –
Ah, – knew he but the Astrologer's Decree –
Take – take me hence! –

They are going.

DE VERE. Nay – Geoffry – Agatha –
I pray explain to me this marvellous Scene? –
Are ye all mute?

They turn back.

AGATHA. She thought you wounded, Sir.
CERULIA. I thought thee slain! – Pardon the witless Joy
Of quick surprize in seeing thee once more;
And take my humble thanks for Geoffry's safety.

Going.

DE VERE. There was no merit; offer me no thanks.
The life I risk'd I wish not to preserve.
CERULIA. Alas, it breaks my heart to hear such phrase!
Ye gracious Powers! – If Hubert be not happy –
Nay, come – we but obtrude –

going.

DE VERE. No, stay thee, maiden.
If there be aught in Hubert's shatter'd power
Frankly command –
CERULIA. Oh, far too generous! – No! –
AGATHA. Unless to Geoffry's Father –

II 310 *No, raise ... head*: an insertion, replacing 'Lean on my friendly arm'.
II 312 *knew he but*: an insertion, replacing 'had he known'.
II 314 *I pray*: an insertion, replacing two variants: 'Some one' and 'Prithee'.
II 315 *Are ... mute*: an insertion, replacing 'Will no one speak'.
II 317 *in seeing*: an insertion, replacing 'to see'.
II 317 *more*: an insertion, replacing 'again'.
II 319 *offer*: an insertion, replacing 'give'.
II 319 *me*: 'then' is deleted.
II 323 *Nay*: an insertion, replacing 'But'.
II 323 *but obtrude*: an insertion, replacing 'trouble him'.
II 324 *shatter'd power*: an insertion, replacing 'power to serve thee'.

GEOFFRY. Ay, true; There
One favouring word might win his hard consent,
And with our vows entwine our gratitude.
DE VERE. Enough. I sleep not till thy suit is crown'd.
Off! – off! – no praise, – lest I withdraw my purpose.
And say, Cerulia, with Truth's modest courage,
Hast not Thou, too, some gentle swain to bless?
CERULIA. To bless? – I have naught but Hate for All I see –
Alas – why say I that? – I mean not Hate –
DE VERE. But, should I portion whom thou could'st approve,
Might not a gentler feeling welcome him?
CERULIA. No, Hubert, no! – Great, noble as thou art,
And poor, and humble, and forlorn as I am;
Hold me not yet at such afflicting distance
To deem my heart a prize for bribery!
DE VERE. Fair, wond'rous pattern of instinctive virtue,
I will offend no more. Speak thy life's plan;
And, be it what it may, command my service.
CERULIA. O happiest moment to Cerulia known!
Deigns thy high mind a Thought – a Wish – a look
For poor Cerulia that is not scorn?
I have no plan; – new-born I feel! – no Fear
My breast shall ruffle; no Hope agitate;
Not even a wish shall break on my repose –
Save only for the noble Hubert's welfare.
Blest, blest be He – and woe will fly Cerulia!
DE VERE. Heav'ns! – is it Here that Virtue takes her Stand?
Virtue, that once I thought – Mark me, Cerulia!
Mark me, – and, henceforth, think of me no more.
I merit not thy generous prejudice:
Turn thy fair thoughts, oh Maid! some happier way;
Quit – and forget – who ill deserves thy care.
CERULIA. Art thou then criminal? – Thou, Hubert, Thou? –
DE VERE. No; but unhappy – even to agony! –
CERULIA. And – is't for that I should forget and leave thee? –
Oh no! – for that my heart would cling the fonder –
Ah! woe is me! – Haste – Haste, dear Agatha, –
Ah, why to Him spoke not the Astrologer! –
Come, Agatha. – Pardon me, gracious Hubert; –
I leave thee, as thou bid'st – but to forget thee –

II 341 *Fair*: an insertion, replacing 'Thou'.
II 343 *command*: an insertion, replacing 'accept'.
II 344–48 *CERULIA ... no Hope agitate*: a pasteover.
II 359 *agony*: an illegible line is deleted.
II 360 *leave*: an insertion, replacing 'quit'.
II 362 *Ah! woe is me! – Haste*: an insertion, replacing 'What am I saying?'.
II 365 *I leave*: an insertion, replacing 'I quit'.

Oh, urge me sooner to forget I live!

Exeunt.

DE VERE (*alone*). What touching Beauty! what affecting softness!
What purity of Innocence and Love!
Bright is the view of Honour thus innate
Glowing untutor'd in the uncultur'd mind.
How happy might I prove with that fair creature
Had my deluded soul not fixt its faith
On its own object of elected worship!
Still are we blest – or curst – but by our fancies!
Repining Man, impatient of controul,
Blind to himself, short-sighted to events,
And darkly ignorant of Cause, Means, Purpose,
Judges of All by that inept criterion,
That slave of Impulse, wayward, sickly Self;
Deems Right, or Wrong, what flatters, or disgusts it,
And balances Creation by its wishes.
Self! – various, partial Self! – Thou Thing of Storms!
Blown by conflicting Passions, struggling Vices,
Wild Projects, vain Desires, – resisting Virtues, –
Or, all-enthralling humours of Occurrence
That, by their fond Dominion, chace alike
Advantage or Destruction; Thing of Storms!
Blown, like the Winds in elemental strife
From adverse points, by whirl encountering whirl: –
I marvel at – I scorn – yet bear thy sway. –

ACT III

A rural Landscape, with high Hills in the back Ground, and the Cottage of DE VERE *in Sight.*

Enter DE MOWBRAY *and* GIFFORD, *meeting.*

GIFFORD. My lord, fair tidings!
DE MOWBRAY. Hah! – hast seiz'd Geralda? –
Secur'd her writings? –
GIFFORD. No; but young Cerulia
Has vanquish'd Hubert.
DE MOWBRAY. Hence, then, every fear!

II 368 *What purity ... Love*: an insertion, replacing 'What artlessness of genuine tenderness'.
II 369 *Bright ... view*: an insertion, replacing 'Fair is the sight'.
II 371 *How*: an insertion, replacing 'Yet'.

All danger's o'er. Geralda will abjure him.
Good Gifford! – skilful! – Rare! – but . . . is't no dream?
Art not deluded?

GIFFORD. On her snowy hand
I saw him plight his faith.

DE MOWBRAY. Exquisite hearing!
Why this is Safety, – Reputation – Life!
Give to Geralda an idea of perfidy
I'll trust her haughty soul, and brave my Stars.
How did this tender Lamb tame the fierce Wolf?

GIFFORD. Thou hast seen the brook that wildly runs it course,
Gliding through every aperture, to form
Wandering and winding, it's meand'ring Channel;
Growing more beautiful for obstacles:
And through it's Water's, bright and clear, displaying
Each variegated Fossil, Shell, and pebble
That, smil'd on by the rays from Phoebus shot
Glitter as Gems of brilliancy refulgent;
So, through her countenance transparent, beams
All that her mind concentrates, Hope, Fear, Love,
Sweetness, Anxiety; – each chacing each
As her Soul's object meets, or quits her view:
Till, as the limpid brook, through mystic mazes,
Never repell'd, attains, at last, the ocean;
Her passion, with her purity and beauty,
Has conquer'd ev'n the stubborn heart of Hubert.

DE MOWBRAY. O, I could clasp the Gipsey to my own! –
Speak on! Where met they? – Gifford! I am thy Slave!

GIFFORD. I saw him soften'd by her tender powers,
And urg'd pursuit; she thought it destiny,
And follow'd as I bid. At break of Day
She rose to trace his footsteps. At her Sight
Displeas'd, he turn'd, and roughly question'd her.
She wept, and would have fled. He grew more gentle;
And then she told the Astrologer's decree.

DE MOWBRAY. He could not sure suspect –

GIFFORD. No; he but smiled;
And said, Fair damsel, had he named some other
Would'st thou not just as kindly have giv'n Ear?

III 8 *is*: 'Reputation' is deleted.
III 12 *Thou hast seen*: an insertion, replacing 'E'en as'.
III 16 *And through . . . displaying*: an insertion, replacing 'Displaying through its clear, bright Waters'.
III 19 *refulgent*: two illegible lines are deleted.
III 24–25 *Till, as the limpid . . . ocean*: Burney replaced these lines by 'Till it at length attains, through mystic mazes', which renders the syntax incorrect.
III 29 *Gifford! . . . Slave*: an insertion, replacing 'Wast thou on the watch'.

Blushing, she cried: O no! – I had then surmiz'd
The Astrologer was false. This mov'd him greatly.
Faltering became his voice, disturb'd his accent.
Fair maid, he cried, from me no joy can spring! –
Yet soon soft vows were interchanged. Cerulia
Appears Felicity personified;
But Hubert sadder looks; fiercer, more gloomy,
From All, but her, he turns with fresh disgust.

DE MOWBRAY. The contract must be ratified. But Beauchamp? –

GIFFORD. He flies to execute your last commands,
And hasten certain vassals to the Isle.
Where Beauty pleads not, Beauchamp may be trusted.

DE MOWBRAY. See to Cerulia now.

Exit GIFFORD.

So far is smooth; –
Smooth on the surface – but within – what tumult
Stirs! Still, like a Snake insidious, Remorse
Lies numb'd and slumbering while Suspence, while conflict
Hang hovering, quivering o'er the Event, but darts
With ruthless gripe, to massacre Success! –
The fear of failure o'er, the awe remov'd
Of coming Shame and Ignominy, the Soul
With fatal retrospection and enquiry,
Flings on the past it's undivided faculties:
And Then – O terrible! – Then, when fair Victory
Should tune each Nerve to Harmony and Triumph
Grim Horrour fills the Scene, and stalks around
Tremendous; wearing forms inexplicable,
Wak'ning a conscious dread in each direction,
And giving to the aching Eye no rest,
The sicken'd heart, no peace! – This way they come –
My Sight is blasted by the view of Hubert –
Happy be Thou, at least, blameless Cerulia!

Exit.

Enter HUBERT DE VERE *and* CERULIA.

DE VERE. But, will thy partial thoughts, thou tender maid,

III 40 *surmiz'd*: an insertion, replacing first 'been sure', second 'perceived'.
III 42 *Faltering ... voice*: an insertion, replacing 'His voice grew faltering'.
III 54 *Still*: an insertion, replacing 'How'.
III 55–56 *while conflict ... but darts*: an insertion, replacing:
hangs o'er
Event, to dart, with pungent pang inflam'd.
III 57 *With*: an insertion, replacing 'And'.
III 59 *Of coming Shame*: an insertion, replacing 'by Shame, Defeat'.

Be permanent? O, I am taught to fear! –
Wilt thou indeed be faithful? – Shall no other –
CERULIA. None before Thee could ever touch my heart; –
Who, after Thee, would try?
DE VERE. 'Tis wond'rous all!
Ah, could I merit such soft excellence! –
CERULIA. Merit the poor Cerulia! – Generous Hubert!
Thou turn'st my glee to tears, talking such phrase!
O! I will watch thee all the livelong day,
Attend thy steps, obey thy varying will;
See when thou woo'st the solitary walk
And instant leave – or from afar regard thee:
Rise with the morning's Sun to range the mead,
And vent my heart's content in Hymns of joy;
Still wait thy bidding ere I seek thy sight
That no Cerulia may unwish'd offend it. –
DE VERE. Ah! thou deserv'st a heart unus'd, – unbroken! –
With all its first fresh feelings, unimpair'd
By Spirit-sinking Time, foul Injury,
And blood-empoisoning, baleful Disappointment! –
Such as I am, howe'er, behold me thine! (*takes her hand*)
And, for the remnant of my varied life
Mould me, Cerulia, to thy own fair will.
CERULIA. Why do I weep? O Eyes, perverse! – ingrate! –
Right noble Hubert, judge me not unkindly;
It is not grief, but bliss too high, too happy –
I breathe with pain through Joy – my heart so swells. –
Would it displease thee should I seek my friends? –
And, when I have told thy gentle grace and favour,
And lighten'd my o'erburthen'd breast – Say, Hubert,
Shall I return?
DE VERE. Do as thou wilt, sweet maid.

Exit CERULIA.

HUBERT DE VERE.

Once more on the adventurous Sea of life

III 76 *Ah, could I*: an insertion, replacing 'Could I but'.
III 90 *blood*: an insertion, replacing 'life'.
III 92 *life*: an insertion, replacing 'days'.
III 93 *own fair*: an insertion, replacing 'gentle'.
III 94 *O*: an illegible word is deleted.
III 96 *bliss*: an insertion, replacing 'joy'.
III 97 *through*: an insertion, replacing 'from'.
III 99 *gentle ... favour*: an insertion, replacing 'goodness – wept and pray'd'.
III 100 *breast*: an insertion, replacing 'bosom'.
III 101 *as*: an insertion, replacing 'what'.

I sail. What can I fear of Storms to come
Compar'd with those I have weather'd? Deep sinks my heart –
Yet, if I make Her happy – gloom may vanish.
Meek Innocent! How swift the roseate glow
Of Hope, discards the lillies from her Cheek!
The love that, unrequited, nipt her bloom
With the pale faded tint of cheerless mildew,
From the first gleam of gentle sympathy
Renews her vivid charm of blushing fragrance.
So see we oft rough Winter's hoary hue
Blown by a northern blast back from its progress
Towards other climes, fasten on Summer's Flowers,
Wither their growth, envellop their sweet odour,
And change their cast to its own dreary aspect:
But when, ere long, the renovating Sun
Breaking through mists, perceives its usurpation,
His quick, resistless beams chace the chill blight
As pleasure's sparks of fire from mental bliss
Turn the dim tear to Beauty's liquid lustre.
Can She prove false? – Perchance! – Traitress Geralda!
Polluting each pure source of confidence
With foul suspicion! What a semblance bore she!
O She had all that here on Earth seems perfect!
Spirit with Softness, Virtue, with fair Wit,
A gaiety to chear the pangs of woe,
A tenderness to calm the storms of ire! –
How may Cerulia – Barbarous retrospection! –
Yet Time, – her worth, – her love, –
A VOICE FROM THE HILL. Help, Stranger, help!
DE VERE. Heavens! – what sounds –
THE SAME. Oh hear me, Stranger, hear me! –

III 106 *swift the roseate glow*: an insertion, replacing 'beauteous'.
III 107 *her*: an insertion, replacing 'thy'.
III 106–07 *Meek ... Cheek*: an insertion, replacing two illegible lines.
III 108 *nipt*: an insertion, replacing an illegible word.
III 108 *bloom*: an insertion, replacing 'sting'.
III 109 *faded tint*: an insertion, replacing 'frost'.
III 109 *cheerless mildew*: an insertion, replacing 'Winter's [an illegible word] light'.
III 111 *charm*: an insertion, replacing 'bloom'.
III 115 *Wither*: an insertion, replacing 'Nip'.
III 115 *their*: an illegible word is deleted.
III 117 *But when ... Sun*: an insertion, replacing an illegible line.
III 118 *Breaking*: an insertion, replacing 'Portent'.
III 121 *lustre*: four lines are deleted:
Thus Love that, unrequited, iced her Spring
With the pale frost of winter, Love itself
With the first gleam of gentle sympathy,
Renews her blushing bloom of young hilarity.
III 128 *calm the storms*: an insertion, replacing 'melt the flint'.

DE VERE. Whence are those accents? – Hah! – What Form –
GERALDA (*from the Hill*). Oh stay! –
DE VERE. What wild illusion strikes my sight? –

GERALDA *darts forward*.

GERALDA. Stranger! –
Whoe'er thou art, oh succour and conceal
An helpless Fugitive, – pursu'd, – oppress'd, –
In pity, in kind mercy, grant me shelter!
Turn not away thine Ear! – Oh list! – Great Heaven! –
That look – those features – ah! – Doubt – Wonder – Hope
Thrill through each vein –
DE VERE. Am I not in a trance?
GERALDA. Oh gracious, pitying, all-consoling Powers!
Whom do I see? –
DE VERE. Burst, burst, my struggling Soul!
Geralda? –
GERALDA. Yes! 'tis he! 'tis he! – Hubert! –
Ah! – I forget myself! – And Thou, De Vere,
Dost Thou forget me too?
DE VERE. No! – Wife of Glanville! –
No! – I am not so blest! –
GERALDA. Alas, surprize,
Uncertainty – alarm – Joy – All conspir'd
To rob my recollection of thy sufferings:
But back they rush with fullest retribution!
Yet, if thy mind retain it's pristine candour,
If Sorrow's shafts, and Injury's corrosions
Have yet not blunted, or consum'd thy feelings,
Repress thy wrath, and hear –
DE VERE. To what avail?
GERALDA. Ah, not to call from thee returning love! –
If There the sluices of thy heart are dry,
Or petrified by solitary rage,
I can but mourn; – I mean not to resist.
'Tis to thy justice that my Fame appeals;
Not to thy passions my affections cry.
DE VERE. What sounds are these? – What phrenzy gives them Ear?
O memory! vengeful guardian of my wrongs
Rise, rise! impregnable in their hard armour!
Oppose thy records 'gainst this new illusion –
Those records that so oft have riv'd my Soul

III 159 *What phrenzy ... Ear*: an insertion, replacing 'Why do I stand to hear them?'.
III 160 *vengeful*: an insertion, replacing 'faithful'.
III 163 *riv'd*: an insertion, replacing an illegible word.

With lacerating wounds, from ties dissolv'd,
From artful Smiles, false vows, pretended virtues,
Betray'd affections – and remember'd blessings!
Remember'd – but to be bewail'd for-ever!

GERALDA. If Memory and Geralda rise together
In Hubert's breast, will they not call back time
To that fair period when no blast assail'd
Geralda's Honour? when her truth –

DE VERE. Hold! hold! –
What is thy aim? would'st thou again delude me?
No, no! – The time is past!

GERALDA. Hubert! – if e'er
I seem'd to thee such as thy praise portray'd
Condemn me not unheard! I am not culpable!

DE VERE. Not culpable? – Where hide ye now your heads
Oh Shame! Oh Truth! Oh lost Integrity!
Not culpable? – Tell me, then, what is Guilt?

GERALDA. By hard compulsion doom'd to wear its semblance,
I marvel not thou hast judg'd me thus severely:
Yet, could'st thou bend thy rage to hear my story –

DE VERE. No, madam, no! play on me now no more!
I stand upon a precipice thou see'st not! –
If thou could'st prove – Egregious supposition! –
If thou couldst prove – The word choaks utterance! –
If, in the widest compass of strange wonders
That circle this amazing Globe, some mystery –
Some dark, incredible phenomenon
Swell in Time's womb to rush upon my senses
With sounds uniting Innocence and Thee –
Avaunt, avaunt – insane imagination!
Avaunt, vain phantom of delirious Terror!
Unequal'd Misery!

GERALDA. Hubert!

DE VERE. Fare thee well –
Why should I trouble thee to vex thy mind
With quaint inventions, marvellous details,
And windings long of plausibility?
No! let me spare the task! –

going.

GERALDA. De Vere! De Vere! –

III 173 *Hubert*: 'De Vere!' is deleted.
III 174 *praise*: an insertion, replacing an illegible word.
III 187 *Globe*: an insertion, replacing 'World'.
III 189 *Swell*: an insertion, replacing 'Hangs'.
III 192 *Terror*: an insertion, replacing 'fright'.
III 193 *Misery*: an insertion, replacing 'Horrour'.

But that I see the tumult of thy soul,
The wild disorder glaring in thy Eyes,
The tremulous motion of thy livid lip
Confessing – all thy tongue disdains to own, –
How could I brook such fierceness? bear such insult?

DE VERE. Right, right! – 'tis best we part. I am too harsh:
I'll leave thee to discourse to yonder hills
Of Innocence and Guilt, and Hubert's roughness.
They'll hear at least in silence – but for me
Discussion comes too late. – My mind's absorb'd –

pauses.

What were to me thy guilt – or innocence? –
Adieu! –

Exit abruptly.

GERALDA.

GERALDA. Away, then, every hope, – or fear! –
And ends all thus? –

Re-Enter DE VERE.

DE VERE. One moment – I return.
Something – I know not what! – I recollect
Imperfectly and vaguely – Didst thou say
Thou wert pursued? – distress'd? – in search of shelter? –
If it be so, my Hut shall give thee refuge,
My Arm – defence.

GERALDA. O noble, even in wrath!
Ev'n with injustice, noble! – yet I ask not
This generous effort Now; I flew pursuers
Who strove to rob me of the Sight of Hubert;
He flies me now himself; – What would they more?
I will no longer shun them.

DE VERE. Whence this mockery?
The sight of Hubert? – Was't not the work of chance?
Could'st thou design to meet thy perjury's victim? –
Feast on his anguish? – Revel in his change? –
His Fall? – his Ire? – his Groans? – The sight of Hubert? –
Is it not baleful to thy Eyes? blighting
To all thy views? Repugnant to thy thoughts?

III 209 s.d. *abruptly*: an insertion, replacing 'precipitately'.
III 223 *Feast on*: an insertion, replacing 'Witness'.
III 223 *anguish*: an insertion, replacing 'misery'.

Death to thy hopes? and to thy Conscience shame?
GERALDA. O! wert thou sick – with Pain, with toil, with peril
Say, had I follow'd thee?
DE VERE. Me? – Follow'd me?
Perfidious woman! art thou not another's?
GERALDA. No, Hubert! –
DE VERE. Hah! –
GERALDA. No, Hubert, no! – The duties
Which sacred rites, howe'er compulsive, claim
Have been fulfill'd, and – Hubert! – now are clos'd!
DE VERE. Hah! – art thou free? –
GERALDA. I am! –
DE VERE. Flash, light'nings, flash!
Consume me with your forked flames! – Here! – Here!

(*Tears open his Garment*)

Leave not an Atom! – Powder me to ashes! –
My Brain – my burning Brain's on fire! – Free, saidst thou,
Free?
GERALDA. Oh whence this rage?
DE VERE. Com'st thou to blast me?
To rob me of the pittance poor remaining
Of sad submission to my evil Stars? –
My evil Stars? – Oh blasphemy! – My evil
Is but thy guilt! – Thy guilt? – true! thou art guilty! –
I breathe, I live again! Guilty, most guilty! –
Reckless I view thy freedom or thy bondage. –
Since thou declin'st my service, peace be with thee: –
I cannot wish thee ill; – Farewell! – Farewell! –

Exit.

GERALDA.

GERALDA. What exquisite distress! His wrath is feeling;
'Tis sensibility too finely quicken'd

III 237 *my burning Brain's on fire*: cf. Mrs Delvile in *Cecilia*, p. 680.

III 228 *Pain ... peril*: an insertion, replacing 'peril, pain, and toil'.
III 231 *Hubert, no! – The duties*: an insertion, replacing 'Thou knowest not, then, the duties'.
III 235–38 *Consume me with ... to blast me*: a pasteover.
III 241 *Oh*: an insertion, replacing 'What'.
III 243 *Guilty, most guilty*: an insertion, replacing 'Reckless, unmov'd'.
III 244 *Reckless I view thy*: an insertion, replacing 'Alike, I view', which in turn replaces an earlier illegible passage ending with 'thy'.

That deviates into wayward irritation.
What mischief has been wrought him, what dread havoc
By brooding rage, and solitary woe!
But Time and Truth – that never long are sunder'd, –
These clouds will clear. What though the blossoms fall
Through rude and angry blasts, the Stem remains
Firm, though despoil'd, and noble, though a wreck,
Till milder Seasons bring it's foliage back,
Gracing it's melancholy dignity
With Autumn's wreaths: so when the stormy passions
That thus uproot his nature's gentler feelings,
Allay'd by Reason, sooth'd by tender love,
Forgo their fierce tempestuous ascendance
De Vere again will rise his Sex's model.

Re-enter DE VERE, *precipitately; his hands crossed over his Forehead.*

DE VERE. My Senses are at War! – Chaos is here! –
Why com'st thou? – Speak! – Unravel thy intent –
Thy secret motives! – If there rest a doubt –
However vague, improbable, – impossible! –
Hovering between thy guilt or innocence
Clear it with light'ning's speed! – To leave it here
Pent in this tortur'd brain – O it might bring me
Somewhat too near distraction! –
GERALDA. Stay, then, thy wrath,
Silent, if not compos'd. De Vere! – thou know'st;
There was a time –
DE VERE. Ay, send it to oblivion! –
If to that time thou turn'st when thinking thee
Spotless as angels whom thy form resembled
I lov'd thee – and believ'd myself belov'd –
GERALDA. Yes, let us pass it o'er; and rather dwell
On that, to thee, it seems, far happier period
Which bearing thee away to distant wars
Postpon'd, at first, and then annull'd a contract
That else had giv'n to thee this hand despis'd
This heart abjur'd and scorn'd. –
DE VERE. Proceed, proceed!

III 261 *Forgo*: an insertion, replacing 'Forfeit'.
III 263–64 *My Senses ... thy intent*: an insertion, replacing two illegible lines.
III 270 *distraction*: 'for thy safety! – ' is deleted.
III 270 *Stay ... wrath*: an insertion, replacing an illegible word followed by 'Wilt thou then calm thy spirits to attend'.
III 271 *De Vere! – thou know'st*: an insertion, replacing one and a half illegible lines.
III 280 *That*: an insertion, replacing 'Which'.

GERALDA. Scarce wert thou gone, when O! – a storm arose
That shook to their foundation all my hopes
And in one mass of ruin buried them!
Glanville, returning from the Court of Philip
With articles of peace 'twixt France and England,
Betray'd a Writing, sign'd by sundry Barons
Yet unenroll'd in treason's open records,
Offering to France this interdicted Kingdom,
By secret treaty, from the unpopular John.
DE VERE. Base, foul expedient! – Oh come the Time
When Glory knits in one our King and Country!
GERALDA. This writing to my uncle he reveal'd –
That guilty uncle! – foremost in the list
Appear'd De Mowbray's name!
DE VERE. Was he a traitor?
How 'scap'd he Glanville's charge?
GERALDA. Ah, how! – alas! –
Glanville, as mede of silence, claim'd Geralda
With half De Mowbray's wealth.
DE VERE. How! – Knew he not
Thou wert already mine? – Barbarous woman!
To thee was that unknown? In Honour mine?
Name, name the tie, the bond, the sophistry
That, cancelling Honour, forms an higher claim?
GERALDA. If Honour dwell in any single point
I stand condemn'd! for I have forfeited
My word, my given faith, my plighted love!
But hear the cruel motive. –
DE VERE. O, I know it!
A proud alliance with a richer husband;
A true, true woman's motive.
GERALDA. Hold, De Vere!
Thy rage I have borne; but . . . try not thy contempt!
With warm resentment, and with cold disdain
Alternate, I repell'd their force united,

III 285 *Philip*: Philip II, King of France.
III 290 *unpopular John*: Hume relates how rebel barons, affronted by John's insolence and tyranny and reduced to extremity by his use of foreign mercenaries against them, 'offered to acknowledge Lewis, the eldest son of Philip, for their sovereign' (*History of England*, I, 450). Louis came to England in May 1216 but abandoned his claim in September 1217.

III 287 *Betray'd*: an insertion, replacing 'Brought forth'.
III 291–92 *Base, foul . . . Country*: an insertion, replacing 'Most foul expedient to each patriot heart'.
III 293 *reveal'd*: an insertion, replacing an illegible word.
III 300–01 *To thee . . . the sophistry*: an insertion, replacing:
Did'st thou not know it too? In Honour mine?
Tell me then, what's the tie, the fraud, the sophistry?

Till, desperate and enrag'd, Glanville fulfill'd
His hardy menace, and to John confess'd
The dark conspiracy.
DE VERE. And yet to Glanville,
This wretch vindictive, thou could'st yield thy hand?
GERALDA. Pierce not a mind already wounded! –
DE VERE. Wounded? –
Is thy mind wounded? – O could my life's blood heal it! –
Bring back sweet peace, and call again fair virtue
There – to their wonted mansion! – to that bosom
Which erst – Go on! –
GERALDA. The treason thus was known
But not it's authours. Still in Glanville's power
De Mowbray's life remain'd. – on me 'twas cast! –
O what dire scenes ensued! what tears, what pleadings!
Prostrate to me, my uncle sued for grace;
Prostrate to Glanville, I implor'd his pity –
DE VERE. Thee prostrate? – Paint not such indignity!
GERALDA. Brief be the rest. The conflict still impended
When John's dread summons from Lord Glanville claim'd
The fatal list of the Conspirators –
Ah, brief indeed! – that hour, that death-like hour –
DE VERE. Hold! – Go no further! –
GERALDA. No! – the tale is told! –
Geralda sav'd the Brother of her Father! –
DE VERE. And was it thus – O! I grow sick at heart! –
Thus undefiled – without a stain or blemish –
I lost thee? –
GERALDA. View these writings, fall'n to me
By Glanville's death.

Holds some papers, which he snatches and tears.

DE VERE. Scatter them to the winds! –
I see thy chrystal Truth, bright, pure, transparent!
Thy radiant excellence, thy blaze of virtues –
See them with agony – with horrour see them –

III 315 *yield*: an insertion, replacing 'give'.
III 322 *'twas cast*: an insertion, replacing 'he cast it'.
III 324–25 *Prostrate ... his pity*: an insertion, replacing:
Prostrate to Glanville, I implor'd his pity,
Prostrate to me, my uncle sued for life – .
III 329 *The*: 'list – the' is deleted.
III 334 *Thus undefiled*: an insertion, replacing 'Could it be thus'.
III 337 *I see thy ... transparent*: an insertion, replacing 'What can they prove? – I see thee as thou art!'.
III 338 *Thy radiant*: an insertion, replacing 'See all thy'.

GERALDA. Is it, then, pain to thee I am not guilty?
DE VERE. Pain! – 'tis perdition! 'tis the heaviest evil
Fate yet has pour'd on this devoted head!
Worse than thy perjury – bitterer than thy falsehood –
Ev'n than thy loss more dire! –
GERALDA. Nay, then, farewell!
If it delight thee to believe me criminal,
If it can soothe thee still to think me false
Delighted be, and sooth'd: no vain appeal
Again shall reach thine Ear from one so scorn'd.

Going.

DE VERE. Scorn'd, didst thou say? – to thee, Geralda, scorn? –
O! could I scorn thee – what terrific torture
Were wrested hence! (*striking his bosom*), what pungent pangs! – what woe
Unheard of! – What excruciating regret!
GERALDA (*returning*). O Hubert! – if these ravings speak a mind
Disorder'd by fell anguish, – not an heart
Unstrung by Nature's dearest harmonies,
How could I yet bear with each wild invective,
Brook thy injustice, deprecate thy hate,
And trust to Time's all renovating power
To clear the mists that cloud thy noble Reason!
DE VERE (*approaching her*). Transcendent goodness! – Art thou still so Kind?
So patient still? Dost thou not deem me now
Grown furious? Savage? aptly qualified
With howling Wolves to herd that haunt the Woods?
But all unfit for such as Thee to mix with?

Retreating.

GERALDA. No, Hubert, no! I deem thee what thou art,
Victim of wrongs that loudly cry for vengeance.
Is it for me, through whom thy sufferings sprang
To hate thee that thou feel'st them? Hubert, no!
If Heaven allow'd to sympathy the power
To wrest those griefs from others that we share
This heart, this pitying heart, would foster thine,
Take them all home, and die to give thee peace!
DE VERE. Where am I? (*wildly*) Transport – Anguish – Bliss – Despair
Oh conflict insupportable – Here end! –
Geralda! Yes, behold me at thy feet! –

III 344 *dire*: an insertion, replacing 'dreadful'.
III 351 *pungent ... woe*: an insertion, replacing 'pangs, what misery'.
III 350–52 *O! ... excruciating regret*: a pasteover.
III 367 *sprang*: an insertion, replacing 'flow'.

Thou generous model of true female greatness!
Thou fairest pattern of exalted worth,
More dear than the first hour of my blest vows! –
Wilt thou endure to hear – to know –

GERALDA. Rise, rise!
Lord of my constant heart! my Hubert, rise!

DE VERE. This were felicity too great for Man
Did not the prize relentless that still weighs
All human bliss in scales with human woe
Bear down even now –

GERALDA. Alas, some step approaches –
I fear pursuit – De Mowbray's treacherous arts
Ev'n since I landed – Ah! conceal, conceal me! –

DE VERE. Behold yon simple Hut, my humble Dwelling;
If thou canst deign –

GERALDA. Hubert, to share Thy fate
Will e'er ennoble mine. Enclose me quickly.

GERALDA *enters hastily into the cottage of* DE VERE, *who shuts the gate, and moves watchfully forward.*

CERULIA (*from within*). My Hubert.

DE VERE. Oh Heaven and Earth! O lost Cerulia!

He flies into the Wood.

Enter CERULIA.

CERULIA. He hears me not! – Ah! – if he lov'd like me! –
How would he watch! how listen to each breeze!
How mark it's playful waving through the foliage
Whispering to every Leaf Some step approaches!
I will not follow; will not strain his Mind
To force a bidden welcome. Could that bless me?
Free be he as the Birds that fly each other,
Yet, circle in their flight to meet anon.
Again I'll seek my Friends, my early Partners.
The Choral Warblers of the shady Grove
Will not pour forth their Nightly melody
In sweeter concord, or in softer sounds
Than the Companions of Cerulia's Grief
Will hail, with generous hymns, Cerulia's Joy.

III 386 *Ah! conceal, conceal me!* – : an insertion, replacing 'Is there no concealment?'.
III 391 *Enter* CERULIA ... *like me*: a pasteover.
III 392 *watch*: an illegible word is deleted.
III 393 *it's*: an insertion, replacing 'their'.
III 397 *that*: an insertion, replacing an illegible word.
III 403 *Grief*: an insertion, replacing 'days'.

ACT IV

The Landscape represents a Country Village.

Enter CERULIA *followed by* AGATHA, PHEBE, AMY, *and a Group of Shepherdesses.*

CERULIA. Come, all ye Village Maids, come! share my joy! –
Shall I not be too blest?
AGATHA. No, sweet Cerulia;
What is the bliss exceeds thy fair desert?
PHEBE. We love thee more and more each setting Sun.
AMY. Yes, dear Cerulia, my own heart beats lighter
To see thee gay as thou wast wont to be.
AGATHA. O, when I view'd thee pensive, wandering, cheerless,
Ev'n by the side of Geoffry I felt sad:
And oft, when dance and song invited sport
Wherefore, he'd cry, my Agatha, this tear?
CERULIA. Kind, dear Companions of my earliest joys!
Soft, pitying soothers of my recent woes!
How may I thank you all? How prove my love?
For thee; my Agatha, the generous Hubert
Ev'n at this hour, perchance, is working good:
My tender Friends! will ye not All delight
His bounteous heart, taking his willing aid?
PHEBE. When thou art wedded, I may whisper thee
Something of Humphry? –
AMY. Shall I whisper too? –
Who knows but, for thy sake, my faithful Colin –
CERULIA. Oh, he will grant whate'er you ask or wish:
Scarce would you know him now, his brow serene
Bends with such sweetness as he holds discourse.
Hard by yon Hawthorn as this Morn we parley'd
How his Eyes soften'd! how those fierce, bright beams
Which, erst, shot fire – and frighten'd all but me –
How were they chang'd to melting melancholy!
Kinder and kinder still, softer and gentler
He grew each moment –
AGATHA. Could he look, or list,
And help to love thee?
CERULIA. Tears of wondering joy
Roll'd down my Cheeks, so quick, so fast, so flowing –
Can it be me, I thought, for whom fair Fortune

Thus gilds with sunny aspect watery clouds?
Can it be me the noble Hubert honours?
Does not my fond delight beguile my senses?
Do not my hopes, my wishes, play me false?
Oh blessing Hopes! – Oh Wishes, so alive
Ye glow as granted boons! – Can you – Oh no! –
You cannot thus elate but to destroy! –
Illumine, but to dye with deeper tint
Some hovering, dark reverse. – No, no, – soft Hopes!
Enchanting Wishes! You but teach my heart,
By soft Expectance, Joy's bright harbinger,
To meet this burst of bliss without delirium!

AGATHA. Twice has the Sun his annual course renew'd
Since here he dwelt, yet, till this morn, I ne'er
Beheld him with thy Eyes – but how he brighten'd
When he my Geoffry nam'd!

PHEBE. Had I been there! –

CERULIA. Why say they that the World is full of cares?
Mine all are past and o'er; and when I am Hubert's
Round and around the village will I ramble,
That not an Eye may weep, a heart know sorrow.
O, what a bliss of bliss will be Cerulia's! –
If you have any griefs, pour, pour them on me
Lest I grow wild with happiness unmixt!

AGATHA. Behold De Vere.

CERULIA. Quick let us glide away
And wait his call.

PHEBE. No, no; 'tis thine to meet him.

CERULIA. Think'st thou he knows not I shall wish his bidding?
O, but this morn I told him all my heart –
Let me not e'er obtrude. Come with me, then,
Come! all ye Village maids, and share my joy!

Exeunt.

Enter HUBERT DE VERE.

DE VERE. She flies me! – though unconscious of the doom
Her flight retards; – flies through soft modesty,
That source divine of female dignity.
How may I mitigate – What can I offer? –
Interest she heeds not; and Ambition knows not;
Her soul, by some unseen fatality,
Is all consign'd to misplac'd tenderness.
In what a labyrinth of distress I have stray'd

IV 54 *pour, pour them on me*: an insertion, replacing 'give me your portion'.
IV 64 *That source divine*: an insertion, replacing 'The pure, first source'.

Through grateful pity! – perhaps Vanity! –
How it all chanc'd – my Senses mock belief! –
O my Geralda! from thy generous Ear
Remote be the dire tale, till the sweet maiden,
Instructed in my sorrows, grants them respite:
Thy Soul, though form'd for genial sympathy,
Though organiz'd for Love and Joy extatic,
As Apathy is torpid; as Misanthropy
Repelling and austere, where Happiness
Would decorate Life's passage with a Floweret,
A sprig, a bud, that is not cull'd by Virtue.
Remote from Thee the tale! – or, sick, oh sick! –
My shuddering heart forebodes Resolves – subversive
Of peace to All, though springing from fair Honour.
Where, in this perilous life, does safety dwell?
Exiled, abandon'd, fugitive from Man,
Here, sullen, have I pass'd my bitter hours
In wrathful loneliness – yet insecure!
Is there no refuge from imperious evil
On Earth? No covert from insidious danger? –

CERULIA *appears at a distance.*

And wears it such a Form? – She bends this way –
Seeking, perchance, the meeting she may rue: –
No! let me hope the blissful power of Youth
Bounding, elastic, o'er the woes, the cares,
The Disappointments fell of Life's vicissitudes,
May ope to new-born views the chang'd, not lost,
Expectance, ever fleeting – of Felicity.
O buoyant, smiling, happy – transient period!
O golden pliancy of juvenile glee!
How vain the search of thy gay, fearless feelings
When thy Spring stiffens into hard maturity! –
She's here! – No, she retreats: – The village Maids
Around her crowd: – Would She were from the throng! –
Cerulia! – Fair Cerulia!

Enter CERULIA, AGATHA, PHEBE, AMY, *and Shepherdesses.*

IV 83 *springing ... Honour*: an insertion, replacing 'by fair Honour modell'd'.
IV 84 *does*: an insertion, replacing 'is'.
IV 84 *dwell*: an insertion, replacing 'placed'.
IV 92 *blissful*: an insertion, replacing 'pristine'.
IV 97 *happy*: an insertion, replacing 'blissful'.
IV 97 *period*: an insertion, replacing 'innocence'.
IV 99 *feelings*: an insertion, replacing 'prospects'.

CERULIA. May I come?
AGATHA. May we All come?
PHEBE. To wish thee, noble Hubert,
Life's fairest blessings with our lov'd Cerulia –
AMY. To tell thee –
DE VERE. O, no more! – Send hence thy Friends. –
PHEBE. May we not hope –
CERULIA. Haste, Haste away, dear Maidens;
Hubert some other time may bid you welcome.
SHEPHERDESSES. All health, all joy, to Hubert and Cerulia.

Exeunt.

DE VERE *and* CERULIA.

DE VERE. What strange, uncouth annoyance! –
CERULIA. Nay, forgive them,
Kind Hubert; they but mean respect and love,
Though in their zeal they grew perchance too bold.
DE VERE. Of them no more: For other themes, Cerulia,
Summon attention. – How may I unfold –
What language find. –
CERULIA. Perhaps I came too soon?
Shall I away awhile? – and then – return?
DE VERE. Thrice lovely virgin! gentle, fair, and good –
Say! – hast thou e'er in fortitude been tried?
CERULIA. O Yes! Two lingering Years I fondly lov'd
Where I could cherish not a hope – a thought –
Save what some Friends imparted. –
DE VERE. Well I know
That all of virtue genuine, pure, innate,
Soul-springing, soul-inspired, is Thine, Cerulia:
But thy soft mind, unpracticed in misfortune,
Suspects not yet the deadly weight of woe
That mortal man can bear. The first event
That wounds our peace, to Inexperience seems
A poignard to cut short the threads of life: –

IV 128 *poignard*: variant of poinard or poniard; dagger.

IV 104 *thee*: an insertion, replacing 'We must not be denied'.
IV 104 *noble Hubert*: 'To give thee, noble Hubert, joy, to wish thee' is deleted.
IV 105 *Life's fairest blessings*: an insertion, replacing 'Each blessing'.
IV 107 *May we not hope*: an insertion, replacing 'Yet let us first'.
IV 107 *Haste*: an insertion, replacing 'Hasten'.
IV 110 *Nay, forgive them*: an insertion, replacing 'Be not angry'.
IV 111 *Kind*: an insertion, replacing 'O'.
IV 111 *mean*: an insertion, replacing 'meant'.
IV 113 *Of them no more*: an insertion, replacing 'We will not speak of them'.
IV 122 *pure, innate*: an insertion, replacing 'pure and instinctive'.

Could'st thou endure, then, –
CERULIA. Ah, what could I not,
O noblest Hubert, for this tender asking?
What canst thou wish I shall not joy to grant?
This heart's first pride is but to please – to serve thee;
So soft, so soothingly thou ne'er hast spoken: –
To win, – to keep thee thus – would I not die?

GERALDA *is entering, but perceiving* CERULIA, *starts and stops.*

DE VERE. Thy goodness rends my heart! – O direful task! –
Fairest Cerulia! summon all thy force
To meet thy fate, and to decide De Vere's.
Know that thou stand'st the all-commanding guard
Of every avenue to my future peace;
Of every solace, each fair hope, –
CERULIA. My Hubert! –
GERALDA. Ah! –

She utters a groan and sinks.

DE VERE. Geralda! – Fainting! – Heavens! – my life! my love! –
Oh inauspicious moment! –

Flies to support her.

CERULIA. Ah! – Where am I?
Art thou distract? – Hubert! – What sayst thou, Hubert? –
Turn! – Speak to me! – Oh turn! – No; help the lady –
Succour, support – but call her not thy love!
DE VERE. Oh complicated misery! – Geralda! –
Soft-beaming Life's suffusion tints thy Cheek –
Raise, raise thy Eyes!

GERALDA *rises, but looks away from* DE VERE, *and leans against a Tree.*

IV 136 *Fairest*: an insertion, replacing 'Sweetest'.
IV 137 *To meet ... De Vere's*: an insertion, replacing 'Know that each avenue to my future peace'.
IV 140 *Of every ... hope*: an insertion, replacing:
All means of solace for my life to come
Are in thy power, on Thee depend.
IV 141 s.d. *sinks*: an insertion, replacing 'falls'.
IV 146 *Succour, support*: an insertion, replacing 'Aid, succour her'.
IV 148 *Soft-beaming ... Cheek*: an insertion, replacing 'Returning Life's suffusion tints thy cheek'.

CERULIA. She trembles . . . Gentle lady,
I'll guide thee to my Cottage. Through yon trees
See'st thou not it's green thatch?
DE VERE. Speak, oh speak! –
A word were life! –
GERALDA (*in a faltering voice*). Leave me, my lord De Vere!
DE VERE. Oh fatal moment! – Oh! –
CERULIA. Come, lady, come! –
Why dost thou gaze thus on her? – Turn not from me! –
Look not away! – Am I not thy Cerulia? –
Thy plighted love? – Hubert! – thy promised Bride?
GERALDA (*waving her hand*). Leave me, my lord De Vere! –
DE VERE. O fair Cerulia,
Haste to thy Cot! – I pray thee. – I conjure –
CERULIA. Alone?
DE VERE. Oh go! –
CERULIA. And quit . . . this stranger lady
So faint? – so sick? –
DE VERE. Hie, hie thee hence, oh maiden! –
CERULIA. Without one gentle word?
DE VERE. O, if thy heart
Be ope to soft compassion – if thy breast
Be tuned to tender woe – hie, Maiden, hence! –
CERULIA. And . . . wilt thou follow?
DE VERE. Haste – in mercy haste!
CERULIA. Shall I e'er say thee nay? – but O! – come quickly! –
I know not where I am! – my senses wander –
Oh come! with Life and Love . . . or, Death and Pity!
Exit slowly, turning frequently back.

GERALDA *looks after her in silence*, DE VERE *covers his Eyes with both his hands.*

GERALDA (*after a pause*). My lord De Vere! – whate'er this scene portends
Geralda's fate and thine henceforth are sever'd.
DE VERE. This is the doom my secret soul has dreaded,
And which to avert –
GERALDA. That secret monitor
Urged it not more? – Urg'd not, this village maid
Whose plaintive voice still vibrates to my heart,

IV 162–63 *Be ope . . . Maiden, hence*: an insertion, replacing:
Be ope to pity – if thy breast be tun'd
To soft compassion, hie thee, maiden, hence!
IV 172 *Urg'd*: an insertion, replacing 'Told'.
IV 172 *Urg'd*: an insertion, replacing 'told'.
IV 173 *heart*: 'strange' is deleted.

Since thou hadst sought her love, deserved thy faith?
DE VERE. I never sought it! by that sacred Honour
Which firm – through all the injuries of fate,
It's trials, it's temptations, and it's woes
Hubert de Vere has held inviolate,
I never sought it – Oh! – stab me not with Doubt! –
GERALDA. I know thy truth and thy existence one –
Yet – what can I infer?
DE VERE. That I am wretched –
Not perjur'd! – Oh thou mistress of my Soul,
From which nor Time, nor Wrongs, nor Treachery
Could blot thy cherish'd memory, – not perjur'd!
Yet would I fain have sav'd thy generous feelings
Availess suffering for the deep distress,
Strange, terrible! – in which I stand involved;
And spar'd vain conflicts for yon fair Unfortunate.
GERALDA. Relieve my wondering expectation.
DE VERE. Mark me!
Worn, shatter'd, and recluse, hating the Day,
Here, from the aera of my banishment,
Slow, wearing, and forlorn, my life has wasted.
Hard by me dwelt, with Eustace, an old Shepherd,
Her Guardian, or her Sire, a lovely Virgin,
Fresh as the ambient gales that wave yon Woods.
Airy and blyth she rov'd the fragrant mead;
Light seem'd her heart, gay as her bounding step.
The new Young Grass, elastic from her foot
Rose, as untrod by mortal weight, and touch'd
But by the fluttering Birds she nimbly chac'd,
Not to annoy, but emulate in speed.
How, or by what impell'd, still moves my wonder,
But ever and anon she watch'd my footsteps.
If in the vale I dragg'd my weary walk,
Seeking retir'd refreshment; there I saw
Skirting some Hedge, beside a Tree, or tracing –
Some Brooke meandering wild, her gentle Form:

IV 176 *firm*: an insertion, replacing 'still'.
IV 179 *Oh ... Doubt*: an insertion, replacing 'Wilt thou now believe me?'.
IV 195 *Fresh ... Woods*: an insertion, replacing 'Fresh as the ambient air of this sweet Dale'.
IV 196 *blyth*: an insertion, replacing 'gay'.
IV 197 *gay as*: an insertion, replacing 'and light'.
IV 199 *untrod*: an insertion, replacing 'untouched'.
IV 206 *a*: an insertion, replacing 'some'.
IV 206 *Skirting ... tracing*: an insertion, replacing 'Between the Trees, beside some bush, or near'.
IV 207 *Brooke*: an insertion, replacing 'stream'.

If on the Hills I woo'd the raging tempest,
Congenial to the tumult of my soul;
This delicate maid would court the assailing storm
Though trembling, tottering from it's rude, bleak blast.
Whether I stroll'd on th'open heath expos'd,
Or through the windings of th'intricate wood
Still 'twas the same; she cross'd me in the path,
Or, first at distance, watch'd my wandering motions.
Yet never spoke she; and her modest Eye
If mine perchance it met; was quick averted;
While the bright mantling on her roseate cheek,
And swift-wing'd, eager flight afar, display'd
Youthful confusion.

GERALDA. What a touching picture!

DE VERE. Long time I thought it chance, nor heeded her; –
Spare the recital how her virgin love
Betray'd it's secret: still it writhes my heart
To think of her illusion. Oh Geralda!
Ne'er till this morn eventful, big with sorrow
From deep regret, yet opening to such transport
That sorrow will not hold its seat; stern Justice
Forbears reproach; and pitying sympathy
Loses its poignancy in grateful bliss –
Ne'er till this wonderous morn my mind gave way! –
O had I sooner known thy truth! – hadst thou
Deign'd me one ray of light! –

GERALDA. Reproach not silence:
The Wife of Glanville – by whatever means, –
Became the Guardian of her husband's honour.

DE VERE. O barbarous chance that hither banish'd me!
O fatal spot! – Why was I exil'd hither?
Wherefore that edict of disgrace –

GERALDA. De Vere,
Thou wilt forgive the fallen? – Geralda's Uncle –
Reflect! should'st thou revenge, we still are parted! –
Thy Name inserted where his own had stood! –
And Glanville, with a rival's jealous hate,
The fraudful deed display'd, which Ruin urg'd,
Yet secresy, by deadly arts, exacted.

IV 208 *on*: an insertion, replacing 'up'.
IV 208 *woo'd the raging tempest*: an insertion, replacing 'sought the rudest wind'.
IV 209 *Congenial ... soul*: an insertion, replacing 'Best suited to the fever of my feelings'.
IV 210–11 *This delicate maid ... rude, bleak blast*: a pasteover.
IV 222 *virgin love*: an insertion, replacing 'partial thoughts'.
IV 223 *Betray'd it's secret*: an insertion, replacing 'Became unravell'd'.
IV 225 *sorrow*: an insertion, replacing 'sadness'.
IV 227 *sorrow*: an insertion, replacing 'sadness'.

DE VERE. Oh! monstrous! –
GERALDA. Glanville, in Death-bed agony,
The double guilt reveal'd, the mutual perjury,
And with his dying hand thy honour clear'd:
Then sunk: the Grave enclos'd his last Remains.
Recluse, I waited to cast off those Weeds
Worn not by Grief, nor Love, nor kind Respect,
But custom's decency, and Christian's mercy;
Then – bade De Mowbray chuse or Flight – or Death! –
And, bearing to this Isle Glanville's confession,
Hither I flew.
DE VERE. O pattern pure of Virtue!
Hear, then –
GERALDA. No! – first Reply! – Has this fair Virgin,
Whose touching innocence, and artless love
Affect, melt, win, – Has she no Right? – no Claim? –
Disorder rises on thy brow! –
DE VERE. Geralda! –
Hear calmly – What not calmly! – I reveal –
O dread avowal! – bitter recollection! –
This fatal morn I plighted her my vows! –
GERALDA. Then, Hubert, thou art her's! No earthly power –
DE VERE. Hold, hold, Geralda! – give not utterance
To sounds that herald phrenzy! – True, I plighted, –
Hapless! my vows – but when she knows my story,
Hears what base arts have sunder'd us –
GERALDA. O Hubert!
Canst thou expect so high a sacrifice?
DE VERE. Yes! She will seize the Right from happy instinct,
With chaste simplicity's impulsive honour;
Not taught by rule, not reason'd into practice,
But the fair Spring of elemental virtue,
Great, though unconscious; humble, yet sublime!
GERALDA. If such thy hope, of what angelic mould
The Matchless Maid is form'd! – De Vere, she is thine!
Nature, in concord sweet of dulcet Harmony
Compos'd you for each other.
DE VERE. Oh Geralda!
Offend not thus thy virtues, nor my faith.

IV 256 *melt, win*: an insertion, replacing 'engage'.
IV 258 *I reveal*: an insertion, replacing 'I confide'.
IV 273–74 *The Matchless ... Harmony*: an insertion, replacing:
Is not she form'd! And canst thou wish to lose her?
No, take her, Hubert! – Nature, in sweet concord.
IV 277 *thoughts*: two lines are deleted:
It's secret wishes, every darling hope
To thee consign'd, in thee fulfill'd, in thee – .

If thou couldst read my soul – it's inmost thoughts, –
GERALDA. And yet – Oh Hubert! – yet –
DE VERE. I thought thee false!
Else, the united World, in vast conspiracy,
In vain had watch'd the wavering, wandering Moment
That might have stolen this breast from thy dominion.
Say, then, no restless fear, no anxious scruple,
Unfounded, vague, perturb'd, again shall part us:
Restore thy noble trust –
GERALDA. Could it be shaken?
No, Hubert, no! see this fair Excellence
Reviv'd, resign'd, serene –
DE VERE. And then? –
GERALDA. And then,
Geralda is thy own!
DE VERE. Noblest – most generous! –
Thus at thy feet –

Enter CERULIA.

CERULIA (*as she enters*). Hubert! Thou com'st not? – Ah! –

DE VERE *rises in confusion*: GERALDA *stands aghast*: CERULIA *seems transfixed; then wrings her hands with convulsive emotion, and runs wildly off*.

DE VERE. Great Heaven!
GERALDA. Oh Hubert! Hubert! – We must part! –
Fly, fly – pursue – o'ertake – console – restore –
DE VERE. Conjuncture dire! – I go; – but Oh! – beware! –
Illusive heroism lights thy Eye
Portentous – Oh beware! – Of Woe to All!

Going.

GERALDA. Hear, Hubert, hear! – New to calamity,
And stranger to the boisterous storms of life,
Will not her timid and affrighted Spirit
Shatter'd, o'erpower'd, sink beneath their weight?
Geralda's lot, severe in Sorrow's school,
Deep has impress'd that hard experience
Whose lessons, though they chill the Spring of Hope,
Rescue, at least, Affliction from Despair:

IV 286 *Reviv'd, resign'd*: an insertion, replacing 'Consoled, reviv'd'.
IV 286 *And then*: 'Hubert! – De Vere! – with Reverence – with Love' is deleted.
IV 288–293 *CERULIA . . . Of Woe to All*: a pasteover.
IV 295 *And*: an insertion, replacing 'A'.

Is it not mine, then, to resume the task,
The discipline, long practic'd, to support
Life with its ills, when it has lost it's charms?
But this sweet Flow'r a Zephyr might destroy:
Raise her, then, Hubert: Solace –

DE VERE. Is it Geralda –
Judgement's sage Arbitress, Reason's pure guide,
Would ward off present pain by future suffering?
Unvers'd in human fate, this gentle maiden
May deem her heart decides her happiness,
And fear no evil but it's disappointment;
Short-sighted to the distant, deeper mischief
Waiting its partial and ill-priz'd success
But – when she saw me wasting with despair;
When she perceiv'd to joy no joy responsive;
When to the voice of Love she heard no echo, –
Could she be happy? No! Gifted by Nature
With Passion's genuine ardour, chastely pure,
Though Born of glowing Sensibility
Would she not soon discern the daily struggle
Of Woe suppress'd, with Happiness assum'd?
Woe vainly stifled, woe innate, obdurate,
Glued to my soul, appendant to my being,
Breathing to barbarous Memory's aching sense
Geralda lost again – and lost for ever!

Enter, hastily, AGATHA, PHEBE, AMY, *and other Shepherdesses.*

AGATHA (*rushing forward*). Say, Hubert, say, what has befall'n Cerulia?
Wildly she flies us All, with look distrait,
And Eyes to Heaven uprear'd; and when we urge
To know her Will, or Wish, she answers, Death!
Death is my only wish!

DE VERE. Immortal Powers!

AGATHA. O, she was made of gentleness and Joy
Ere this morn's wily snares bewilder'd her.
Take back your grandeur – it has blasted her;
Take back your grandeur, and restore Cerulia!

DE VERE. Peace, peace! – Where may I find her? Point the Way.
O'th'heights? – I'th'vale? –

AGATHA. Alas, she darted on
With winged speed, charging we would not mark her.

IV 318–19 *With Passion's ... glowing Sensibility*: an insertion, replacing:
With all its glowing sensibilities
Array'd in chaste, untutor'd innocence.

GERALDA. O fly, De Vere!
DE VERE. I am gone. This lady, Agatha,
Guide to thy Cottage – (*to* GERALDA) O Adieu! – (*to* AGATHA) Tell Geoffry
That with his life he answers for her safety.

Exit precipitately.

GERALDA. Dread interval! Ye gentle Damsels, speed me
To your abode; and, as we pass, relate
Something of this fair Virgin.
AGATHA. She is our pride;
Our village glory; darling of the plain:
The Flower whose bloom spreads health; whose fragrance, joy,
But Oh! whose blight would nip each hope i'th'blossom,
And wither every heart.
GERALDA. Alas, poor Maid!

Exeunt.

Enter DE MOWBRAY *and* GIFFORD.

GIFFORD. Fear not detection; scarce Geralda's self
Could know thee in this garb.
DE MOWBRAY. And what avails
The transient respite? Ruin scowls before me: –
They have met! There needs no more! – I prithee, Gifford
Is there no bark? – no skiff? – no fishing boat
Can waft me to the Continent? I care not
How, nor which way.
GIFFORD. Is all so desperate? – Beauchamp,
With three stout vassals, lands ev'n while we parley.
DE MOWBRAY. Hah! – Then one effort yet remains – That past –
Yet what have I to lose? – Hie thee to Beauchamp.
Charge him to stay the Vessel. – Could ye sieze
Geralda – Gifford, all might yet be wrested
From fierce destruction, though its flashing fires
Scorch me all o'er.
GIFFORD. My lord, it shall be done.
But – Will not Hubert follow?
DE MOWBRAY. Aye; no doubt;
I mean no other. Fly! – secure the Vessel! –

Exit GIFFORD.

Come, there is yet one chance to snatch at! – O!

IV 352 ***bark … skiff***: both 'bark' and 'skiff' are terms for a small boat.

IV 355 ***lands … parley***: an insertion, replacing 'has just reach'd the Isle'.

I'll grapple for it, till each limb disjointed
Sink torpid from the purposes of life.
If Hubert follow – I have friends abroad;
If here he rest, – Cerulia shall enthrall him.
Beauchamp I dare not trust – nor dare discard! –
O, 'tis the curse avenging Rectitude
Has stampt on Frailty, that one erring step
Impels another, and 'gainst all mankind
Lurking distrust awakens. On we plunge, –
Guilt is Guilt's herald, Crime, Crime's harbinger!
Why hate I Hubert but that I have wrong'd him?
Why through each vein thrills Rancour's baneful venom,
Geralda, at thy Name, but that my projects
Make thee, perforce, the Martyr to my safety? –
Abhorr'd Reflection! with what subtlety
Glidest thou thus in, 'twixt all that I can plan,
Or act, or think; black'ning each fairer prospect
Each rising hope! No peace, no rest, no quarter
Giv'st thou my wearied Soul from Memory's probe.
O let me once more grasp security,
Once know what 'tis to breathe unfearfully,
And I will study pennance Morn and Night
To soothe this troublous Conscience. How it gnaws me!
O ye – who have not practic'd wilful wrong
How little do you know of misery!
De Vere – Geralda – ye are blest to me! –
Blest ev'n Thou, too, Cerulia – though cast off –
Friendless – alone – abandoned – Still, still blest! –

Enter GIFFORD.

GIFFORD. My lord, this moment favours every wish:
Beauchamp secures the Vessel, and awaits
Your orders; while, in yon lone hut, defenceless,
Lady Geralda lies conceal'd.
DE MOWBRAY. Once more
Fortune, I then defy thee! villainous Hag! –
Most honest Gifford! – Be she seiz'd o'th'instant –
But keep thou still with me; the ways – the paths
Unknown, I else might stray. – Yet – how defenceless?
Where is De Vere?
GIFFORD. Some evil has o'erta'en

IV 369 *nor dare discard*: an insertion, replacing 'yet must employ'.
IV 373 *Lurking*: an insertion, replacing 'Constant'.
IV 384 *more grasp*: an insertion, replacing 'but taste'.
IV 400 *Yet*: an insertion, replacing 'But'.

The young Cerulia, who, with wild dismay
Sudden has disappear'd.
DE MOWBRAY. Cerulia, saidst thou? –
Malignant Fate! What curse must menace next?
GIFFORD. Alarm as by contagion through the village
Extends it's ravage. Damsel after Damsel
Weeping, I have met. The consternation spreads;
De Vere himself, forsaking even Geralda,
With frantic terrour roves the country round.
DE MOWBRAY. O tenfold curse should misery fasten There!
Thou demon Fortune! reck not There thy vengeance –
There, where alone some feeble, weak atonement –
Spare me that blow! O in thy mighty rage
Confound not suffering innocence – Why star'st Thou? –
I speak but to the air; – Perchance I dream; –
For there are dreams in hot Imaginations
That wait not for the drowsy lull of slumber
To paint wild visions, to present strange objects,
To work the passions, crucify the Nerves,
And set the Brain on fire! – Fly! – hasten Beauchamp; –
Convey her to the Vessel; – Hubert next
Shall occupy our thoughts. – This peasant Girl –
Think'st thou she's safe? – Go on! – Why sleepst thou thus? –
O Life! – laborious, toilsome – durst I crush thee! –

ACT V

The Scene represents a Country Church Yard.

GEOFFRY *and* AGATHA, *meeting.*

AGATHA. O Geoffry, bringst thou to my heart no solace?
No tidings of Cerulia?
GEOFFRY. Agatha,
Solace I hop'd from thee!
AGATHA. Oh, whither fled? –
Oh lov'd and lost! where hast thou past the Night
The dreary, wasting Night, of length eternal?

IV 408 *De Vere*: an insertion, replacing 'Hubert'.
IV 415 *I speak ... I dream*: an insertion, replacing 'I am talking to myself – dreaming perchance'.
IV 418 *paint*: an insertion, replacing 'give'.
IV 424 *crush*: an insertion, replacing 'end'.

GEOFFRY. No spot I have left unlook'd around the Village;
No face I have met, but all bedew'd with tears:
And, turn which way I would, the Bird of Night
Lowering and Squalid, sail'd before my Eyes,
Shrieking discordant sounds of horrid omen.
AGATHA. My Geoffry, there is yet another woe;
Was evil ever single? That high Dame
Left to our charge so fondly by De Vere,
While we were absent All, away has vanish'd.
I fear those cruel Men – yet if, perchance,
This wandering lady should be Hubert's love? –
GEOFFRY. Lost were we then; Hubert would turn our foe.
AGATHA. His Favour will but blast us if he is false.
'Twill play foul mildew on our plighted troth,
Gnawing its freshness, and it's bloom corrupting.

Enter HUMPHRY.

HUMPHRY. Joy, joy to Agatha!
AGATHA. Where is Cerulia? –
All else is grief! – Is there no trace? – no hope?
HUMPHRY. Across the Church Yard glance thy Eye –
AGATHA. Cerulia! –

Exit, running.

HUMPHRY. Nay, stay thee, Geoffry. Would'st pursue the Winds?
Her maiden Friends, encircling, buz around her
Like Bees that swarm to save their honey'd store,
And, labour as you may, you will not pass them.
They tell me she looks wild, bereft of reason;
Binding her brow with Cypress; – yet, still gentle.
GEOFFRY. 'Tis wond'rous where so long she may have rov'd.
De Vere all Night has wander'd in her search. –
HUMPHRY. She comes! Behold the group!
GEOFFRY. I'll hie to Hubert.
This gladsome news may blunt the lady's loss.

Exeunt.

CERULIA *enters, pale and faint, her head bound with wreaths of Cypress; her Hair disshevelled and flowing*; AGATHA, PHEBE, AMY, *and other Shepherdesses, follow, or surround her.*

V 29 *Cypress*: a light transparent material resembling lawn or crepe; sometimes used as a kerchief for the head in sign of mourning.

V 18 *Favour will*: an insertion, replacing two illegible words.
V 33 s.d. *her head ... wreaths*: an insertion, replacing 'followed by Agatha'.
V 33 s.d. *and*: 'the' is deleted.

CERULIA. Come, all ye village Maids, come, once again!
But not to share my joy! – my joy is flown!
Come, all ye village maids! – oh come once more –
Yet not to share my woe – woe is my heritage.
AGATHA. Oh my Cerulia dear, how have we sought thee,
Trod o'er thy favourite paths, bid Eccho call thee, –
CERULIA. Vain was thy search; far from all Mortal Sight
My doom I waited.
AGATHA. Waited thy doom?
PHEBE. Oh where?
CERULIA. Have ye not heard my sad reverse of fate?
Know ye not, then, my lost, illusive hopes?
Alas! I had thought the World a bloomy Garden,
Planted with smiles, that shone like sun-beams round me,
Than Roses sweeter, and as snowdrops pure.
Now I see clearer! 'tis a wilderness,
With here and there a gleam of fertile culture
To lure the unwary Traveller from his road;
While all beside is thorny, miry, rugged,
Winding aloft to fearful precipice,
Where whoso, Witless! climbs to hail the scene
Is headlong hurl'd in some affrighting Gulph! –
Hubert is false! –
AGATHA. Oh base and barbarous change!
PHEBE. Why dost thou look so scar'd? – so pale? – so alter'd? –
AGATHA. Why hast thou fled us all?
AMY. Where wast thou hidden?
CERULIA. My date is near! My hour is almost come! –
Weep not, dear Maidens!
AMY. Death is in her face!
AGATHA. O woe is me! – O die not, sweet Cerulia!
CERULIA. Last Night! – last Night! –
AGATHA. Forsake us not, Cerulia!

V 37 *woe ... heritage*: an insertion, replacing 'that were fresh sorrow'.
V 39 *bid ... thee*: an insertion, replacing 'call'd thee aloud'.
V 40 *far ... Sight*: an insertion, replacing 'my Agatha, all vain'.
V 41 *My doom I waited*: an insertion, replacing 'Far from each mortal Eye, my doom I waited'.
V 41 *Oh where*: a line is deleted:
AMY. And why? And how?
PHEBE. And why?
V 44 *bloomy*: an insertion, replacing 'blooming'.
V 46 *snowdrops*: an insertion, replacing 'violets'.
V 47 *Now I see*: an insertion, replacing two or three illegible words, which originally replaced 'Now I see'.
V 54 *barbarous change*: two lines are deleted:
PHEBE. Hubert? Can Hubert prove false-hearted?
AGATHA. Why dost thou look so strange, so scar'd, so alter'd, so [an illegible word], so pale?

CERULIA. 'Tis past! – 'tis fixt! – Like some poor fragile Weed
By rude or wanton hand unkindly sever'd
From the tall prop that seem'd to give it strength,
Yet but sustain'd its weakness, Hubert lost
Gone is my hold on life: faint, faded, drooping,
I shake – I fall! – Last Night! – last Night! –
AGATHA. Alas,
I heard the Death Bell toll! –
AMY. Heaven seem'd on fire! –
AGATHA. I thought myself pursued! –
PHEBE. Tell of last Night.
Where couldst thou be conceal'd?
CERULIA. Lo There! – The Church! –
ALL. The Church! –
AMY. Most terrible! –
AGATHA. I fear to breathe!
PHEBE. How found'st thou entrance?
AGATHA. Oh forbear!
CERULIA. Thou knowst
The mouldering Window near the Thorn bush? There
I spied an opening; so my fate ordain'd.
Unseen, I glided in; – unheard, unmark'd.
PHEBE. Did not strange sights appal thee?
CERULIA. Murmuring sounds
And distant echos sad, broke from the Earth;
While from the vaulted roof shrieks and shrill cries
Pierc'd to my brain: and Beings wing'd and wild
Oft cross'd my way; now circling, now pursuing,
Till, spent, o'erpower'd, motionless I fell,
Calling on Death!
AGATHA. Oh, roll not thus thy Eyes!
CERULIA. Long in dread silence, watching some portent,
Shuddering I lay; chill'd, frozen, bloodless. –
Clos'd were my aching Eyes, and yet I slept not;
Hush'd were my Sighs – I fear'd to hear their moan;
And Hour chac'd Hour, in awful mute expectance:
At length – appeared a vision! – white it's garb –
AGATHA. Cerulia, cease! –
PHEBE. What utter'd it? –

V 82–83 *Long in ... bloodless*: an insertion, replacing:
Long in dread silence, shifting, numb'd, I lay,
Watching some token; praying, shuddering, chill.
V 86 *And*: 'awful' is deleted, as is an illegible word which it originally replaced.
V 86 *awful mute*: an insertion, replacing an illegible word.
V 87 *vision*: 'all in' is deleted.
V 88 *AGATHA*: 'I am terrour-struck' is deleted.
V 88 *What utter'd it*: 'Oh tell' is deleted.

AMY. Oh tell! –
CERULIA. That deep even in the bowels of the Earth
These feeble hands must fashion my last home.
ALL. Oh Heaven!
PHEBE. How did it look?
CERULIA. Meagre, yet stern;
Holding a taper whose red flame illumined
Three hideous Spectres that before me glar'd.
And thrice the torch was quench'd, and all was dark,
All deadly gloom: – all sound, all sight denied:
And thrice, by touch unseen, the taper blaz'd
Furious afresh, wav'd by the vision gaunt;
While sulphurous sparks of fire whirl'd o'er my head.
Come to thy task! It cried; thy work awaits thee!
AGATHA. Oh, how I tremble! –
AMY. Hist! – What sound is that? –

They all listen.

AGATHA. Fear'st thou t'offend it, shouldst thou more reveal?
CERULIA. Submiss I rose, obedient to my call.
Twelve struck the Clock; and every Monument
Ecchoed each stroke, in groans that died in murmurs.
Softly I trod; and oft with fear retreated,
For at each step started a flitting Ghost
To wave me to the appointed spot of obsequy.
Thrice There I knelt; thrice call'd on Hubert's name –
AGATHA. Oh faithless Hubert!
AMY. Oh, Deceiver fell!
CERULIA. Haste, haste, I cried, Cerulia 'tis invokes thee!
Grant thy lost victim her funereal rites:
Come, Hubert's Spirit, come! prepare the Abyss! –
Leave her not here, to Worms, to Crows, to Vultures –
And yet, poor Animals! –
They will but feast upon her Mortal clay; –

V 92 *taper*: 'red' is deleted.
V 100 *sound*: an insertion, replacing 'noise'.
V 101 *more reveal*: an insertion, replacing 'tell us more'.
V 102 *Submiss*: an insertion, replacing 'Shivering'.
V 104 *murmurs*: 'Sick look'd the moon, that faintly light me on' is deleted.
V 106 *Ghost*: 'passed by' is deleted.
V 110 *Cerulia ... thee*: an insertion, replacing 'to poor Cerulia's tomb'.
V 111 *thy*: 'poor' is deleted.
V 111 *rites*: 'Hast thou not kill'd and wilt thou not inter her?' is deleted.
V 113 *to Vultures*: an insertion, replacing 'and Ravens'.
V 114 *poor*: two illegible words are deleted.
V 115 *feast*: an insertion, replacing 'fasten'.

They will not fasten on her fairy wishes,
Consume her peace, devour her Spring of life –
Feed on her virgin hopes, and tender love! –
Ye hungry Worms! Ye famish'd Birds of prey!
Are ye All Kinder than the noble Hubert?
AGATHA. O Day of woe! –
CERULIA. Next, – so the vision bid,
The stated space I mark'd, and hail'd my Grave!
AGATHA. Oh cease, Oh cease!
PHEBE. Say on, oh sweet Cerulia!
CERULIA. Prostrate I fell; eager, I bared my breast,
And with the fresh damp mould I strew'd it o'er:
Thrice then I cried aloud: Inter this First!
Come, Hubert's spirit, come! Thou hast broke my heart,
Come, give it Sepulture!
PHEBE. Did he appear?
CERULIA. Sudden a flash of light –
AGATHA. No more! no more!
Thy look is wild! –
CERULIA. I shriekt – I knelt – I prayed! –
Till, faint the dew of Death crept, clammy, o'er me
And Icicled my Heart: and There, even now,
Cold, cold it clings.
AGATHA. Oh most affrighting vision.

V 116 *wishes*: 'Consume her opening youth [an illegible word], feed on her genial bliss' is deleted.
V 117 *Consume her*: an insertion, replacing an illegible word followed by 'her heart's'.
V 117 *her Spring*: an insertion, replacing 'her life'.
V 118 *her*: 'gentle' is deleted.
V 118 *tender love*: 'Soft Hope and gentle Joy and tender love' is deleted.
V 121 *woe*: a line is deleted:
PHEBE. Go on –
AMY. Proceed!
V 122 *The stated ... Grave*: an insertion, replacing 'I measur'd out full length the stated space'.
V 123 *Oh cease, Oh cease*: two lines are deleted:
And, quick from Head to Foot I marked my Grave –
My [two illegible words] and no more! Suspend I – suspend –.
V 123 *sweet Cerulia*: an insertion, replacing 'Speak, sweet Cerulia, speak – ' and an illegible line.
V 124 *fell*: an illegible word is deleted.
V 124 *Prostrate ... breast*: an insertion, replacing an illegible line which begins with 'Prostrate' and ends with 'fell'. Two alternative lines are also deleted:
CERULIA. Prostrate I fell, trembling, Eager I bar'd my breast,
On the cold Earth, and there I bar'd my breast.
V 129 *light* – : two and a half lines are deleted:
but ask no more
What next I saw I ne'er must tell! – I rose, –
I shriekt – I knelt – I pray'd – for Hubert pray'd! – .
V 131 *faint the dew*: an insertion, replacing 'the faint dew'.

Where, where is Hubert? – He may love thee still;
Perchance some errour –
CERULIA. No! – he loves me not!
I have seen the peerless lady of his Choice;
High is her port; graceful, and like his own –
Oh vain Cerulia, that could hope to win him! –
The luckless Moth was I, with gaysome wings
Circling my Sun; and oft, in playful rounds,
Adventurous, scorch'd; Now, by genial warmth
Allur'd, advancing; Now, by dazzling light
Abash'd, retreating; but still hovering nigh,
Till giddy, heedless, darting, tumbling, fluttering –
I reach'd the flame – to perish in its blaze! –
I am faint! – thy arm! –
AGATHA. Lean, lean on me! – Fly, Phebe,
For Hubert –
CERULIA. No; support me to his presence.
And oh, be quick! – pale Death comes striding towards me! –
See you him not? – Look at his ghastly Form! –
Hear you him not? – List to his dreadful yell! –
AGATHA. Where? – Where? –
PHEBE. Say where? –
CERULIA (*kneeling*). Oh spirit of Release! –
Grant yet one moment for my last farewell!
Give me to view, and bless, and humbly pardon,
That no ungenial thought, struggling for strife,
May rive my parting breast, and mar its peace! –
He is gone! – My prayer is heard!

She rises.

PHEBE. Where? – which way went he?
CERULIA. No mortal Eye can mark the path of Death.
'Tis here – 'tis there – it circles Me, and You! –
It winds about us; it involves our footsteps;
No Labyrinth has a clue but, at each outlet
Abrupt he starts to grasp us! –
AGATHA. Dost thou rave?
CERULIA. The Sun is early set . . . So are my Joys!
The Moon is on the wane – So are my Sorrows!

V 144 *darting*: an insertion, replacing 'springing'.
V 144 *Till giddy, heedless, . . . fluttering*: an insertion, replacing 'Trembling, giddy, simple, springing, fluttering', which in turn replaced 'Trembling, fluttering, springing, giddy, heedless'.
V 146 *Lean . . . Phebe*: an insertion, replacing 'Fly Phebe, fly for Hubert'.
V 156 *Where*: an insertion, replacing 'Which'.

And never, Thou gay Sun! shall these dim Eyes
Hail thy bright rays again: No more, Thou Moon!
Thy silver hue congenial shall inspire
The lonely trust that sooths the secret woe! –
On, – on, – I sink! –
AGATHA. Help, Phebe! –
AMY. Sweet Cerulia!
Exeunt, leading out and supporting CERULIA.

Enter DE MOWBRAY *and* GIFFORD.

GIFFORD. Yes, 'tis Cerulia! Whence she comes –
DE MOWBRAY. No matter;
I see her safe: There . . . There, my mind's at rest;
But still – curs'd Fortune's spite! – this adverse Wind –
GIFFORD. No Vessel can oppose it's mighty fury;
But it's first change shall waft to distant shores
Geralda; who, ev'n now, captur'd, conceal'd,
Is from De Vere secur'd as though already
Sever'd by other Climes.
DE MOWBRAY. Still I doubt Beauchamp: –
I am ill at ease; – Would they were gone! – Who's there? –
GIFFORD. De Vere.
DE MOWBRAY. De Vere? – Detain him while I pass! –
I loath his Sight! –
Exit hastily.

GIFFORD. Detain De Vere! – yet hold! –
I'll probe – or pique his honour for Cerulia.

Enter DE VERE.

DE VERE. Soul-harrowing toil! – Where next – Hah! Gifford? – Say
Whether the lovely Virgin of this Village
Missing last Night, has yet appear'd?
GIFFORD. Report,
Pendant in air, for every Wind to turn it
As Malice, Interest, Wit, or Calumny
May suit some purpose, or indulge some passion,
Visits her flight with motives so discordant
That, while Detraction revels in surmize –
DE VERE. Detraction? – Hah! –
GIFFORD. And Envy's hollow smile

V 179 *Sight*: 'I'll not encounter him – Use any feint' is deleted.
V 179 *Detain . . . hold*: an insertion, replacing 'Enough, I know him well'.
V 181 *Where next . . . Say*: an insertion, replacing 'Ha! Gifford knowst thou'.
V 189 *DE VERE*: three illegible words are deleted.

Denotes insidious triumph o'er frail Beauty –
DE VERE. What triumph? – Who is frail? – Of Whom discourse Ye?
GIFFORD. Blushing Timidity and weeping Pity
Bewail her fall.
DE VERE. Her fall? – By Heaven –
GIFFORD. Assertion,
Though of small weight or force in origin,
Soon swells in sound, and bloats in circumstance
Where Mystery generates Rumour.
DE VERE. On my life
If Slander's venom'd, cancerous breath pollute
That sacred Shrine – By Heaven, not my own Honour
Shall urge a deeper, blacker, direr vengeance! –
GIFFORD. Suspence is fertile –
DE VERE. If thou hast tidings
Speak! – If not, – begone!
GIFFORD. Report, my lord, –
DE VERE (*furiously*). Away!

Exit GIFFORD.

Ye roaring Lyons – howling Wolves – fell Tygers –
Devouring Vultures – grinning, fierce Hyenas –
Ye honest Enemies! Instinctive Foes!
I'd court, I'd clasp Ye to the breast ye'd mangle
With Famine's gripe – ere trust the Human Monster
Who, – Hous'd, fed, cherish'd, – unannoy'd, untempted,
With gaiety of Vice, airy though callous,
Sports, wanton, with the purest Gem that Heaven
Has cast on our protection, virgin Fame! –
Why loiters he? – Design sits on his brow. –
How may I pass him to the Cot of Agatha? –
Oh for the respite of one genial instant
With Thee, Geralda, Thee! – In Thy lov'd presence
An Instant has a charm, a Word the power
A look, a Glance the force, – a Smile the witchery
To still – or Wake! – a hurricane in the Soul. –
Yet lurking? – Dare he watch me? – Unreviv'd
Must I renew this fearful chace? – Cerulia! –
Oh should some desperate purpose – Heaven and Earth –
What dire surprize o'erspread her beauteous face –
What sudden woe – Oh! were I curst
With conscious Memory of deluding her
With wily courtesies, to wind about
Her secret heart, impelling trembling Hopes,

V 194 *force*: an insertion, replacing 'worth'.
V 217 *To still – or Wake*: an insertion, replacing 'To wake or still'.

And timid Fears; exciting downcast blushes
That, through the very struggle to conceal
Betray the breast's emotion; – had I practic'd
The ensnaring flattery of persuasive softness,
The silent-pleading smiles of apt attentions,
And all Love's subtle, dangerous Artillery
Which, without claiming, wins the Virgin heart –
Reason would fail me at this perilous juncture. –
The way seems clear. – Ha!

Enter GEOFFRY.

DE VERE. Com'st thou from Agatha. –
GEOFFRY. The lady's gone!
DE VERE. Gone? –
GEOFFRY. And no one knows whither.
DE VERE. Miscreant, 'tis false! – If thou'rt a Villain, Geoffry,
It is but for a moment! – By this arm,
If thou reveal not this accursed plot,
Thy forfeit life –
GEOFFRY. I am no villain, Sir! –
DE VERE. Good Geoffry, no! – I do believe thee honest –
Honest, and Kind, and true! – I crave thy pardon! –
Unfold, then, – Speak! –
GEOFFRY. By all unseen, she vanish'd
Ere poor Cerulia was descried
DE VERE. Descried? –
Has she appear'd? – Thou givst me new existence! –
GEOFFRY. She comes this way. –
DE VERE. Watch her with tenderest care, –
I must be gone – Oh for some clue, good Geoffry! –
GEOFFRY. The Ruffians of last day, perchance –
DE VERE. Distraction! –
Where's Humphry? – Call each Labourer of the district –
Scour all the Country round! – I'll to the Coast, –
Be quick.

going.

GEOFFRY. Cerulia sees thee, and approaches.
DE VERE (*starting*). Great Heav'n, how chang'd! – Oh Flower untimely cropt!

V 229 *ensnaring*: an insertion, replacing 'ensuing'.
V 230 *attentions*: 'Endearing manners, gentle looks, kind accents' is deleted.
V 234 *Com'st thou from*: an insertion, replacing 'Guide me to'.
V 236 *Miscreant ... If*: an insertion, replacing 'Impossible! – If'.
V 242 *By all unseen*: an insertion, replacing 'Unknown to me'.
V 248 *Where's*: an insertion, replacing 'Where'.

Where is thy glowing bloom? – Thy Form elastic? –
Thy airy movement? – Thy celestial grace? –

Enter CERULIA, *leaning upon* AGATHA, *and followed by* PHEBE, AMY, *and Shepherdesses.*

CERULIA. Come, Hubert, come! and take my long Adieu!
Take the sad parting prayers of lost Cerulia. –
(*pointing to the Church Yard*) Behold where lies her Grave! There! – where thy Spirit
Mark'd out the spot.
DE VERE. Oh, ever gentle virgin,
Rive not my heart with woe intolerable! –
Turn back thy thoughts to life! –
CERULIA. It may not be! –
Yet, in thy Memory let me sometimes live.
And, sometimes, when thou wanderest by yon stream,
Treading sweet violets as thou wind'st thy way,
Call to thy Mind There, first, thou saw'st Cerulia;
Stranger to care, unconscious of her doom! –
And cry, Alas, poor Maid!
DE VERE. Heart-writhing sounds! –
(*To* AGATHA) How has this change been wrought?
AGATHA. Alas, last Night
In a dread vision, you appear'd! –
DE VERE. A vision? –
Oh cruel sport of Grief to assail her Reason
With fancied images of aerial horrour!
Sooth her, ye maidens, with your softest skill:
I must away. – A call imperious –
AMY. And can'st thou, Hubert, leave her?
PHEBE. Barbarous Hubert!
Why hast thou given her, why, this deadly blow?
What had she done but love thee? Black requital!
CERULIA. Close to the path that leads to yon high hill

V 253 s.d.–57 *Enter* CERULIA . . *ever gentle virgin*: a pasteover.
V 266 *To*: an insertion, replacing 'Speak'.
V 266 *Alas*: an insertion, replacing 'She saw'.
V 266 *Night*: 'a vision' is deleted.
V 267 *DE VERE*: four or five illegible words are deleted.
V 268 *Reason*: 'With images of [an illegible word] fantastic [two illegible words]' is deleted.
V 270 *Sooth her . . . skill*: an insertion, replacing 'Take her in charge, ye maidens, sooth, support her'.
V 272 *And can'st . . . leave her*: an insertion, replacing 'Hast thou the heart to leave her Hubert?'.
V 272 *PHEBE*: 'cruel' is deleted.

An aged oak o'ershadows a deep glade;
Steep the descent; Oh glide it gently down
In some lone moment of relenting love:
Then pace thee round the wood that borders it,
And thou wilt see, grav'd on the smooth, fair bark
Of many a tall young Beech, thy much lov'd name: –
Will not remembrance bring me then to view?
Though from thy sight I am gone – far off! far off! –

DE VERE. Oh! – What to this were death? – Terrific crisis! –
How go? – How stay? – Kind phrenzy! sieze me wholly! –

CERULIA. Then, shouldst thou seek the Church yard's midnight gloom,
Silent to ponder on Life's fleeting hour;
There, while deep musing, and absorb'd in thought
On weightiest matter, suiting thy high bearing,
Sudden the Moon may cast a glimmering light
On a small stone, beneath the old broken Yew;
Then shalt thou read: Here rest Cerulia's ashes!
Wilt thou not sigh when the last scene is o'er?
Wilt thou not start when slow her death-bell tolls?
Will not a tear bedew her cold remains?

DE VERE. Oh conflict past endurance! – Live, yet live,
Thou wrong'd – unhappy – sweet Cerulia!

Takes her hand.

CERULIA. Oh, ever noble Hubert! – Ah! –

She sinks upon a Bank in the arms of AGATHA.

GERALDA (*from within*). De Vere! –

GERALDA *rushes in, attended by* BEAUCHAMP. DE VERE *darts forward to meet her.*

Once more, escaped – restor'd – once more I fly
To Thee, my lord! my life! – This man of pity –
Thou dost not joy? – Thy look appall'd –

DE VERE. Geralda! –

V 300 *cold*: 'last' is deleted.
V 298–300 *Wilt thou not sigh ... cold remains*: an insertion, replacing an illegible line followed by:
Will not a tear bedew her cold remains?
Wilt thou not sigh when the last scene is o'er?
Wilt thou not start when slow her death bell tolls?
V 303 s.d. *She sinks upon a Bank*: an insertion, replacing 'She faints'.
V 303 s.d. *to meet her*: '*Cerulia faints*' is deleted.
V 304–06 *Once more, escaped ... DE VERE. Geralda!*: a pasteover.

Thine is my Heart! – (*pointing to* CERULIA) But lo! – where Justice calls! –
Oh claims to tear in twain my tortur'd breast! –
If with this life – Calamity's mere sport! –
Her peace . . . and Thine –

GERALDA. Ah! baleful, tragic sight!
Oh all ye powers of Honour – Pity – Equity –
Break, break my heart! – or, make it all your own!

Approaches CERULIA.

Oh canker Disappointment! – could thy blast
Fall only There, to wither in its bud
This bloomy Flower, ere yet its fragrance oped!
No! – Raise her, Hubert, snatch her from the tomb!
Her innocence, her sufferings, her love
Subdue ev'n Me! – What, then, must be Thy feelings?
Save her, Oh Hubert! – 'Tis Geralda's cry!
Geralda's voice shall hail – Oh Heaven! – thy choice,
And pour forth blessings on thy virtuous Nuptials.

DE VERE. Oh my Geralda – noblest of thy Sex! –
Call her, then, back to life! –

Retiring.

GERALDA (*bending over her*). Look up, sweet Maid!
No Rival shocks thy soft, indulgent Eyes,
A tender, pitying Friend they meet, – if Pity
Fade not beneath the reverence of thy Worth.
Look up, sweet Maid! receive the rightful mede
Thy gifted taste, thy high, congenial soul
Bade thee aspire at with instinctive courage;
Uncurb'd by lowliness of fortune; conscious
That in thy breast there dwells a soaring spirit
To meet the noble attributes thou lov'dst,
And claim that grand sublimity of Justice
That spurns all chasm between Rank and Virtue:
Look up, sweet Maid! – Hubert! I can no more!

DE VERE (*returning*). Oh moment of Despair! – Yes, sweet Cerulia!

he takes her hand.

GERALDA. Hubert, farewell! – Oh Hubert! – fare thee well! –
Sustain thy courage – love thy beauteous Bride –
Forget Geralda – or remember her

V 314 *bud*: an insertion, replacing 'bloom'.

But as the deadly source whose poisonous influence
Tainted each shaft that smites thy mangled Breast
With blackest injury, and Grief appalling!

going.

Enter DE MOWBRAY, *with Three men armed.* DE MOWBRAY *points to* GERALDA, *whom they seize.*

GERALDA. Oh Heaven!
DE VERE. Villains, my life goes first!
DE MOWBRAY. Away!
GERALDA. My Uncle!
DE VERE. Oh Fiend, art thou so near?
DE MOWBRAY. Avaunt!

A scuffle. BEAUCHAMP *assists* DE VERE, *who disarms one of the men, and stands upon the defensive against the upraised arm of* DE MOWBRAY, *while* GERALDA *forces herself between them.*

GERALDA. De Vere! – De Mowbray! Oh look there! Respect
This fated victim! –
DE MOWBRAY. Victim? – What victim? – Hah!
Who is't lies there? – Speak! – Answer me! – what horrour
Play ye upon my senses? – Is't Cerulia? –
Speak to me, Maiden!

advancing towards her.

DE VERE (*detaining him*). Hold, thou barbarous Ruffian!
A moment wait: – then, hear – and Answer Me!
DE MOWBRAY (*striking his forehead*). Oh villain! villain! execrable villain!
Oh wretch accurs'd! thus slain in thy own toils! –
Stand off, ye feeble mourners! – All make way
Ye sycophants in grief, weeping for shew. –
What can Ye care? – Oh then wrong'd, murther'd Girl!
'Tis mine to pay thy funeral obsequies, –
And with my blood – my life – my Reputation –
My Infamy! . . . say whether she be dead? –
If she yet live, away! avaunt! begone! –
DE VERE (*guarding* CERULIA). Restrain this frantic rage.
AGATHA. Soft! – soft! – She moves? –
CERULIA. Oh sleep of Death! come! – close these aching Eyes!
Kind sleep of Death – Ah, deem me not too young;

V 344 *My Uncle*: 'DE MOWBRAY' is deleted.

If brief my date in Years, in woe I am aged.
My pilgrimage is o'er; – my Heart's entomb'd; –
No one knows where; – nor how. – I pardon All!
Take me, sweet Peace! – I see thy Snowy Wings –
They wave – they float – Fly not away without me! –
Waft me to purer skies, and Kinder Worlds!
To skies where guileful hopes no more delude,
To Worlds where sorrow breaks the heart no more!

She sinks again.

DE MOWBRAY. Impossible to bear! Look up, Cerulia! –
DE VERE. Now by my life – loosen my arm, Geralda! –
I can no longer grapple with endurance.
Monster, come forth! with every crime polluted,
Thou traitor to thy King, thy Name, thy Country!
Thou Spoiler of my Fame through whose black arts
My injur'd Honour, and my tortur'd Love
Have driven me on this rock of guilt and woe,
Thy blood, – or mine – must bathe this Victim's Tomb.
DE MOWBRAY. Mine is the blood it craves; I bare my bosom.
Fetch but one drop, – the Fiends will do the rest.
I feel them at their work! – They gnaw my vitals! –
Come! Strike! – Thou can'st not wound me. All within
Is slash'd, and hack'd, and slaughter'd, far beyond
Thy puny power, thy mortal instrument,
Thy human vengeance. Come! Strike! Give me Death!
DE VERE. And deem'st thou Death to Thee will herald peace?

They fight. DE VERE *disarms* DE MOWBRAY.

DE MOWBRAY. Off! – off! I mock thy sword! –

struggling to regain his Weapon.

AGATHA. Ah, silence! silence!

DE MOWBRAY *starts*; DE VERE *stands suspended.*

V 364 *My pilgrimage is o'er*: an insertion, replacing five or six illegible words.
V 366 *Take me ... Snowy Wings*: an insertion, replacing 'Come, gentle Death – See, see that pallid Form'.
V 367 *They wave ... without me*: an insertion, replacing 'Those livid Wings! – They wave – they float – Haste, haste!'.
V 387 *herald*: an insertion, replacing 'bring but'.

CERULIA. Oh Sleep, sweet sleep of Death! – why wilt not come?
Sole Friend of woe! – hear, hear an Orphan's cry!
No Father's smile foster'd my opening days;
No Mother's votive Flowers will sooth my Manes:
Him who I lov'd – I have lost – ah Heaven! – my Hubert? –
Flits he before my Eyes? – Hubert! – He is gone –
Forever gone! – Oh dreary, deadly Void! –
Hubert! – all still! – all dark! – Oh soft release!
Oh balmy sleep of Death, come, come – in pity! –

Dies.

AGATHA. Alas, she breathes no more!

Weeps over her.

GERALDA. How soft an exit!
DE MOWBRAY. Not breathes? – Eternal fury be my portion
If thus she die! – Cerulia! – Hovering Spirit!
Hear me, oh hear! I did not mean thy murther!
De Vere – Geralda, were my destin'd prey,
Not Thou – Oh barbarous Chance! – not Thou, Cerulia! –
DE VERE (*throwing away his sword*). Exquisite Justice! is thy hour then come!
DE MOWBRAY. Cast not away thy sword – I court its edge!
Self-lost, o'erwhelm'd – wrest me from public shame! –
Oh vengeful Fate! what have I not endur'd –
Inflicted – forfeited – to avert Disgrace! –
Come, strike! – Thou hast been my tool! 'twas I who wrought
Thy exile to this spot; – bade her allure thee; –
Hah! – art thou mov'd? – Strike, then! – I am her Murderer!
Would'st thou hear more? – Cerulia, wake! – and curse me!
Loud lift thy voice! Reveal the impious tale,
Proclaim to these weak Infidels in crime

V 392 *Manes*: spirit or 'shade'.

V 390 *Sole Friend . . . cry*: an insertion, replacing 'The only friend of woe, an Orphan sues thee'.
V 392 *Manes*: 'No kindred tears bedew the odorous tribute' is deleted.
V 393–97 *Him who . . . in pity!*: an insertion, replacing:
[an illegible word] in their bloom, now grieve to see them fade
Him whom I lov'd I have lost, – Alone, abandon'd;
Oh soft release! tenderest, kindest mercy.
Hubert! – all still! – all dark
Oh balmy sleep of Death, come! – come, in pity!
V 413 *Reveal . . . tale*: an insertion, replacing 'Tell all thou hast learnt'.
V 414 *crime*: an insertion, replacing an illegible word.

Thy murderer is – thy Father! –
GERALDA. Horrour! – Horrour! –
DE VERE. Tremendous Powers! –
DE MOWBRAY. Oh blasted every way! –
A Traitor's death? – No! – by Cerulia's corse –

Trumpet.

Enter a King's MESSENGER, *with a party of Soldiers.*

MESSENGER. Hubert de Vere, and Thou, the Lord De Mowbray,
John, our most potent Liege, gives me in charge
Safe to the Royal presence –
DE MOWBRAY. Ha! – betray'd?
Fury! –

He endeavours to seize a weapon from one of the Guards.

MESSENGER. Secure, and force him to the Vessel.
DE MOWBRAY (*struggling*). Oh for the means to strangle, – stifle, – massacre –
Arrest the vital current – crush these limbs –
Dash out these dastard Brains –

The soldiers force him off.

MESSENGER. Noble De Vere!
The King has been much mov'd by varying rumours
Touching De Mowbray's secret disappearance
Hard on the Death of Glanville. Doubts arise
Of Fraud – of Perjury, – but Fame refulgent
Waits thy avenging call, Noble De Vere.
DE VERE. Fame hangs not on cotemporary breath.
How facile, else, it's blight by Falsehood, Treachery,
And each base satellite of living Envy!
De Vere commits himself to other Judges,
And other times than these; yet I obey;

V 415 *is*: an insertion, replacing 'was'.
V 417 *corse*: four lines are deleted:
By the pale spectre of her injur'd Mother,
By each curst fraud that won – that broke her Heart
This moment is my last!

He aims to stab himself; DE VERE *forces the weapon from him.*

Off! give me Death!
DE VERE. And think'st thou Death – to Thee will bring back Peace?

And, for my country's peace, repress my murmurs:
That peace more sacred to each patriot breast
Than All – save Honour! – to which Man is heir.
(*approaching* CERULIA) Cerulia! – Oh! hapless victim at
Misfortune's shrine!
Thy wrongs, thy woes – with pangs unutterable
Lie bleeding at my heart. – Matchless Geralda!
With what noble pity thy Eye beams lustrous!
Celestial omen! Sympathy divine! –
Oh! weep with me o'er this untimely bier!

GERALDA. Sweet Shade! Ah! See, how gentle still it's look!
How free from all vain cares, all harrowing passions
The calm profound fix'd on it's lovely Brow!
Ambition, Honours, Power, Glory, Rank
What are ye, say! oppos'd to spotless Innocence.
At the dread hour of Life's last scene tremendous?
There, peace serene companion is with Death;
There, Earth's eternal close is void of terror;
Soft the still sigh; resign'd the parting tear;
And bright the Hope that springs to Immortality!

They bend over CERULIA *on Each side; the Village Maids group, weeping, around: and the curtain drops to the sound of the Church Death Bell.*

V 437 *save*: an insertion, replacing 'but'.
V 442 *Sympathy divine*: an insertion, replacing 'Sympathy, Divine Sympathy'.
V 443 *Oh*: an insertion, replacing an illegible word.
V 450 *with Death*: '*They bend over her on each side; the Village maids group, weeping, around, and the curtain drops to the sound of the Church Death Bell*' is deleted.

The Siege of Pevensey

A Tragedy

By the Author of *Evelina* and *Cecilia*

THE SIEGE OF PEVENSEY

There is only one surviving manuscript of *The Siege of Pevensey*, a fair copy in Burney's hand in the Berg Collection.[a] It consists of 240 pages sewn together in five notebooks, one for each act. The writing is exceptionally clear and legible, and there are very few deletions or revisions. The play has not been previously published.

At the end of each scene, Burney indicated the number of lines occupied. Much later, in about 1836,[b] she noted at the end of each act the time taken for a reading: 25 minutes each for Acts I, II, and III, 20 minutes for Act IV, and 25 minutes for Act V. The total of two hours would make the play 26 minutes shorter than *Hubert De Vere* and considerably shorter than *Edwy and Elgiva*, which reviewers found inordinately long (see above, pp. 5, 93). At the end of Act III is a note in d'Arblay's hand, criticizing Burney's decision to term the play a tragedy: 'Il me semble qu'on ne peut appeler *tragedie* une Piece dans la quelle il n'y a pas une goutte de sang versée'. D'Arblay was not, he continued, familiar with the historical source for the play, and thus could not judge to what extent the text 'peut se plier aux changements que je crois qu'il est indispensable de faire'. Since the heroine, Adela, is not a historical personage, Burney was not, in fact, constrained by the need to follow historical sources; the decision to avoid the spilling of blood and to furnish the play with a happy ending was hers.

Burney began writing *The Siege of Pevensey* some months after completing the first draft of *Edwy and Elgiva* and beginning *Hubert De Vere* in August 1790. In an unpublished diary entry, she refers to her work in progress: 'Three works which I have now in hand seize me capriciously; but I never reprove them; I give the play into their own direction, & am sufficiently thankful, in this wearing waste of existence, for being so seized at all'.[c] *Edwy and Elgiva* was still 'in hand', since it existed in draft form only; *Hubert De Vere* and *The Siege of Pevensey* were in earlier stages of development. On 5 June 1791, when Burney began writing *Elberta*, she refers again to three unfinished tragedies, 'not one read! and one of them, indeed, only generally sketched as to plan and character' (*Diary and Letters*, IV, 478). The one 'only generally sketched' was *The Siege of Pevensey*, which Burney, in an unpublished passage, describes as 'in a

[a] The normally reliable *Journals and Letters* errs in stating that two drafts of the play survive (I, 74, n. 6).

[b] See Hemlow, 'Fanny Burney: Playwright', p. 180.

[c] Quoted by Hemlow, 'Fanny Burney: Playwright', p. 180, n. 34.

style my dear friends all around will most wish me to cultivate: but that was not dismal enough' (Berg ms. 4327). In a journal entry of October 1791, Burney refers to 'going on with her third Tragedy', while the two earlier 'tales of Woe', *Edwy and Elgiva* and *Hubert De Vere*, were complete in draft form (*Journals and Letters*, I, 74).

Since there are no further references to *The Siege of Pevensey* in Burney's letters or journals, the date when she made the fair copy cannot be determined.[a] She seems to have valued it less highly than her two previous tragedies, and there is no indication that she ever attempted to have it produced or published.

[a] Hemlow's suggestion that *The Siege of Pevensey* was 'probably completed in 1791 in Dr. Burney's study at Chelsea College' is speculative ('Fanny Burney: Playwright', p. 180).

PERSONS OF THE DRAMA

WILLIAM RUFUS, king of England.
ROBERT, EARL OF MORTAIGN, uncle to the king.
ROGER, EARL OF ARUNDEL AND SHREWSBURY, Governor of Pevensey Castle.
HUGH, EARL OF CHESTER, commander of the king's Troops.
WILLIAM DE WARRENNE, Lord Chief Justiciary.
ROBERT DE BELESME, son to the Earl of Arundel.
RALPH FLAMBARD, Prime Minister of the king.
DE ALDERAY, MOWBRAY, } officers in the service of the Earl of Mortaign.
FITZ HAMMON, DE LACEY, BAINARD, } officers in the king's service.
OFFICERS. SOLDIERS. ATTENDANTS.

ADELA, Daughter to the Earl of Chester.

Prime: an insertion.
ALDERAY: the ms. reads 'Alderi'.

ACT I

Scene i

The Tent of the EARL OF CHESTER. *The* EARL *alone.*

Enter FITZ HAMMON.

CHESTER. Still from the Citadel no news? – no hope
Of bloodless victory? – of quick surrender?
FITZ HAMMON. Said you the Citadel? your tardy speech
Is much behind your thoughts; you mean your Daughter.
CHESTER. I dare not name her! – but the news? –
FITZ HAMMON. There's none:
Earl Mortaign will not yield; and his resistance
Kindles fresh flames each Hour in William's breast.
CHESTER. Of Man's community, and social life,
Is not Ambition the severest foe?
How ample the Dominion of each Brother,
Yet each how ill content!
FITZ HAMMON. Their mutual claim
To seize the Crown, has here a mutual plea:
Robert asserts his right as elder born
Of England's mighty Conqueror; and William
Builds his pre-eminence on that Conqueror's will.
CHESTER. O that the brighter light of liberal justice
Would burst through these dense mists of shallow policy!
Then would they see, that large extent of Empire
Is but large labour; and that never yet
Did Man devise a task so refractory
As ruling Man, – a Creature prone to struggle
With restless skill, 'gainst every thing – but self!
There, all facility, he yields supine
To his each wish, slave of his own caprice,

I i 1 *the Citadel*: Pevensey Castle in East Sussex, four miles north-east of Eastbourne. Pevensey was then a flourishing harbour; William the Conqueror landed there in 1066. It is now, as the sea has receded, a mile inland.
I i 13–15 *Robert ... Conqueror's will*: although Robert, as eldest son of William the Conqueror, would normally have inherited the English throne on the death of his father in 1087, William bequeathed it instead to Robert's younger brother William Rufus, leaving Normandy to Robert.

Though Tyrant of his race.
FITZ HAMMON. Had not Earl Mortaign
So early rank'd himself with William's foes
'Twas his, allied to both, to reconcile
Our England's King with Normandy's brave Duke.
But Robert has his Heart; and from yon Citadel,
Beseig'd as yet in vain, he fights his Cause,
And vows to fight relentless, till the Duke,
His elder Nephew, sits on England's Throne.
CHESTER. The barbarous policy of civil War
Might make the savage blush at polish'd states,
Who forfeits not, like us, experienc'd good.
He ne'er has known the sweets of social compacts,
Nor joy'd in harmony, nor glow'd with patriotism;
Uncultur'd, unconnected, for himself
Alone he lives, and fights, or flies his foe.
Back to those dreary times falls civil war!
Each bond dissolv'd, each genial tie o'er-leapt,
Order revers'd, and Nature violated, –
And all for what? – inordinate Ambition! –
O Times of guilt and woe! Friends assault Friends,
And Brethren, Brethren, – sons commit Paricide, –
And Father's wield the ponderous Arms of Death
Against the sacred Walls that screen their Children! –
FITZ HAMMON. Ah! there the ill comes Home!
CHESTER. Alas, Fitz Hammon,
How vain with craft to struggle against Nature!
There, where the attack is made, – resides my Child! –
There, where I am doom'd to strike, my Heart is lock'd!
There, where I am fated to demolish all
Breathes all for which my life and hope survive!
FITZ HAMMON. Might not the chief Justiciary, De Warrenne,
Work on the king, in Adela's behalf,
To mitigate the order against ransoms?
I know him well inclin'd to plead your cause.
CHESTER. Fitz Hammon, true; but know you not his motive?
No trivial recompense would pay De Warrenne;
My Daughter and her Dowry, – nothing less.
FITZ HAMMON. Is he not rich? And high in William's favour?
CHESTER. What! shall I ransom only to enthral her?
To chain her to an harder, longer bondage?
A life's captivity? she loves him not:
Shall I then sacrifice her mental liberty

I i 25–28 *Earl Mortaign ... brave Duke*: as uncle to William Rufus and Robert, the Earl of Mortaign could have reconciled the two brothers and their respective kingdoms, England and Normandy; instead, he chose to support Robert in his attempt to acquire the English throne.

For the mere name of freedom to her person?
FITZ HAMMON. Were William generous, he had ne'er commanded
A Father's service in this cruel seige.
Perhaps he fears some tampering with your feelings,
And dreads to trust you from his sight in arms.
CHESTER. He knows within those Walls my Treasure's hoarded,
And therefore deems me lukewarm: but he wrongs me.
My guide is Honour, though my Heart is Adela's.
And he shall see I hold my country's cause,
Howe'er less dear, more sacred than my own.
Thus am I driven to double Vigilance,
Fiercer hostility and inhumanity,
Lest – what alone I prize above my Child –
I fail in Duty to my King and Country!
FITZ HAMMON. She yet is safe; in honourable hands
She's fall'n: the Earl of Shrewsbury and Arundel –
CHESTER. No; Robert de Belesme, his eldest son,
Surpris'd her escort as she cross'd the country.
Scarce had he landed from the Norman court,
Scarce trod this shore, ere she became his prize.
'Tis now three months, since lost to my protection –

Scene ii

CHESTER. FITZ HAMMON. A SOLDIER.

SOLDIER. The Lord De Warrenne craves your private Ear.
CHESTER. Give him admittance.

Exit SOLDIER.

FITZ HAMMON. Let me make him way.

Exit.

Scene iii

CHESTER. DE WARRENNE.

DE WARRENNE. With views of friendliness, I haste, my lord,
To inform you that the king has stopt each avenue
By which, henceforth, provision or relief
Might reach the Citadel.
CHESTER. 'Tis well! –
DE WARRENNE. I fear
The bitterest evils now await –

CHESTER. 'Tis warlike! –
Such keen distress may punish and reduce them.
'Tis retribution fit for insurrection.
DE WARRENNE. Yet – the fair Adela –
CHESTER. Ah! speak not of her? –
Who is on guard? –
DE WARRENNE. Bainard. Canst thou endure –
CHESTER. I must perforce endure the chance of War.
DE WARRENNE. Pardon my zeal, Earl Chester, – but your Daughter –
CHESTER. My lord, I know her lost! – Oh mighty Heaven! –
If 'tis by Famine we beseige the Castle
She, useless, will the first be left unnourish'd, –
She, Captive, will the soonest be abus'd! –
Let me not talk of this! – Where is the King?
DE WARRENNE. Surveying the intrenchments. Still, my lord,
Rejected in my suit, I still retain
An interest for the Maid; and with the King
I think I could obtain a dispensation
In this peculiar case, to ransom her.
If 'tis your wish to save her from destruction
Employ me: the last havoc of a seige,
The dangers personal –
CHESTER. Call them not forth! –
I must not think of them! – yet take my thanks, –
O! I would ransom her with half my Earldom! –
If to your influence such a boon were granted
De Chester would be bound to you for life.
DE WARRENNE. Enough! I'll seek the King. –
CHESTER. Yet hold, my lord! –
I do not promise for my Adela; –
Her filial duty merits not compulsion.
What gratitude may work for this deliverance
I know not; but her mind is all susceptible,
And open to each generous propensity.
DE WARRENNE. I'll trust to the event; and lose no time.
Exit.

Scene iv

CHESTER *alone.*

CHESTER. O let me save her from the pangs of Famine, –
– The cries of violence, – and then, resign'd,
Commit her destination to high Heaven!

I iv 1 *O let me*: 'Let me but' is written above the line, but 'O let me' is not deleted.

Scene v

CHESTER. DE LACEY.

DE LACEY. My lord, two persons, muffled and disguis'd,
Sought in the dead of night to pass the guard.
One of them is a Woman. Both are seiz'd.
The man, on some pretence of special business,
Demands an Audience.
CHESTER. See that he is not arm'd,
And let him enter.
DE LACEY. May he not be a spy?
CHESTER. I rather think him some poor fugitive,
Driv'n by the civil tumults from his home:
Else werefore with a female? Bring him in.

Exit DE LACEY.

Scene vi

CHESTER *alone*.

CHESTER. Mournful the Times! we need not make them fiercer
By doubt that acts like certainty. To judge,
We first should try: 'tis easy to be harsher,
But difficult to heal the wounded mind
By sharp severity unjustly pierc'd.

Scene vii

CHESTER. DE LACEY. DE BELESME *in disguise*. SOLDIERS.

DE LACEY. My lord, we found him well supplied with arms,
And scarce from his companion could we part him.
CHESTER. Who art thou?
DE BELESME. One not us'd to be thus question'd.
CHESTER. I fear thy errand will not bear a scrutiny.
What brings thee to the camp?
DE BELESME. A deed of honour.
CHESTER. Thou mayst esteem thy office, because dangerous,

I vii 1 s.d. *in disguise*: an insertion.

Upright; I know the palliatives of policy.
The arts of war admit such sophystry
As in the times of peace would make men shudder.
Serv'st thou Duke Robert?
DE BELESME. See me, Earl, alone,
Then what I am, my purport, my companion,
All shall be told.
CHESTER. In my own mansion, stranger,
I might comply; but here –
DE BELESME. You deem me, then,
Some base assassin?
CHESTER. When man turns foe to man
'Tis not alone humanity he forfeits,
Which ought to link us in fraternal bonds
Impervious to the fierce unnatural strife,
But, further yet, he forfeits common honesty;
Broaches such doctrines, and supports such principles
As drive off native and plain-dealing honour
For wily frauds, the crooked machinations
Of selfish, hard, unfeeling policy.
DE BELESME. I but entreat your private Ear one moment.
CHESTER. One moment be it, then.
DE LACEY. Nay, not alone: –
Much I suspect him of some dark design.
DE BELESME. Hear me but, Earl, and I shall need no Champion.
Call my associate; see us both at once.
DE LACEY. Sir art thou mad? or only ignorant?
Thinkst thou two strangers, taken in a camp –
DE BELESME. Send for her in! Have you not seiz'd my arms?
Whom can I hurt? – or do you fear a woman?
CHESTER (*disturbed*). Well, – let her come.

Exit a SOLDIER.

DE BELESME. Will you not clear the Tent?
Where are my weapons? what my means of mischief?
CHESTER. All is not obvious. War and stratagem
Are kindred terms that join to wound morality.
The rules of nations, as they touch each other,
Might teach barbarity to beasts of prey.
DE BELESME. My follower, like myself, is bound to secresy:
Niether will speak, save singly to Earl Chester.
CHESTER. Then, good De Lacey, watch without the Tent,
And, at the least alarm, rush in.
DE BELESME. Be't so.

Exeunt DE LACEY *and* SOLDIERS.

I vii 32 *disturbed*: an insertion.

Scene viii

CHESTER. DE BELESME.

DE BELESME. And now, Earl Chester, ring the loud alarum
At thy own time. Behold my fair accomplice.

Scene ix

CHESTER. DE BELESME. ADELA, *disguised*.

DE BELESME. Now, potent Earl, reap now the well-earn'd Harvest
Of courteous confidence. Knowst thou this lady?

ADELA *lifts up a large loose hood, and casts herself at* CHESTER'S *feet*.

ADELA. My Father!
CHESTER. Gracious Heaven!
ADELA. My dearest Father!
CHESTER. My Adela! – My Child! – My captive Daughter! –
Do not my senses fail? – My Child! My Child!
Art thou restor'd to me?
ADELA. O yes! O yes!
Again at these lov'd feet thy Adela
Thy benediction craves.
CHESTER. O rise, my child!
Come to my arms! – my Heart!
ADELA (*rising*). O with what joy
Return I to this dear, best, first protection!
O in these honour'd arms couldst thou enclose
Thy Adela for-ever!
CHESTER. How has thou 'scap'd?
How hast thou far'd? – Great Heaven! I dread to ask
Thy treatment?
ADELA. O most noble! Sir, this Warrior –
I cannot speak his praise, – be't thine to thank him.
If Adela is happy in thy love,
If this blest Hour, that sweetly gives her back
O best of Fathers! to thy care parental,
Be priz'd by thee, as 'tis to her most precious –
Look, and behold where Gratitude is due!
See her preserver from the ills of bondage,
See her Deliverer from unheard of horrours!

CHESTER. Brave stranger knight! O be thou whom thou mayst,
Come to this Heart, and share a Father's blessings!
DE BELESME. I could not brook to see such helpless Beauty
Consign'd to all the dangers of a seige,
And risking all the chances of it's crimes, –
It's famish'd groans, it's agonising views
Of slaughter'd Friends, or of triumphant foes, –
Nameless offences to humanity! –
Ev'n Victory, that might have rescu'd her,
Has terrors to bereave her of her senses: –
'Tis Anarchy! 'tis outrage! 'tis Destruction!
CHESTER. How fair is courage, vested thus in feeling!
Now at last speak! Who art thou, gallant stranger?
By whom am I impell'd to love my foe?
DE BELESME. 'Tis Robert De Belesme who stands before thee.
CHESTER. Son of the Earl of Arundel and Shrewsbury? –
Is't possible? –
DE BELESME. The same.
CHESTER. Right noble Enemy!
How will my Heart henceforth recoil to hear
Thy danger or defeat, – late its first hope!
DE BELESME. Myself have brought her; – at the risk of life; –
Of more, O Earl, than life, – the risk of Honour! –
Yet to no other hand could I commit
The fair deposit.
CHESTER. Valourous De Belesme,
What payment –
DE BELESME. None. To see such happiness
O'er pays in joy the rigourous self-command
That urg'd the restoration.
CHESTER. Nay, speak her ransom. –
What-ever I possess –
DE BELESME. Earl, I will speak
When I can find aught within human reach
Worthy to prize with this illustrious maid: –
Till then, I pray thee take her ransomless.
CHESTER. O where are words – Excellent De Belesme! –
I press not on thy honour with my wishes, –
It were un-warrior-like: yet, wert thou, Knight, –
O wert thou with the King! – with Royal William –
DE BELESME. Ah! press not now indeed! now less than ever.
For Eloquence would wear a garb so fair –
Look so like Beauty – seem personified
In a bright Form so unresembling argument, –
Ev'n my own senses, if that way they turn'd,
I must henceforth distrust. – No more of this! –
CHESTER. True, gallant Youth, it suits not with this meeting.

DE BELESME. Now then I go! To linger might be hazardous. –
All Good betide thee, Earl! – save such prosperity
As to my Prince and Cause were woe and mischief.
For Thee, sweet maid –
ADELA. Adieu! –
DE BELESME. For Thee, perchance,
And thy fair lot, I may permit some wishes
That ill accord with enmity! – though now
I owe thee, gentle Adela, much wrong! –
How will my leisure miss thy soft society!
How shall I pass the Hours by contrast ruin'd!
How drag the leaden intervals of War
Which seem'd, with Thee, but countless moments, snatch'd
Indulgent, from the clang of fierce hostility! –
Farewell! –
ADELA. Farewell! – I will not speak my thanks
Lest they detain thee. To the Castle haste! –
DE BELESME. Heaven's mildest blessings light upon thy Head,
Clear, fair, and pure! – I pray thee, Earl, one word. –
When War is o'er, when Blood and Violence
Satiate, have play'd their parts, – if then thy Adela –
This lovely maid – O pardon! – I have done! –
Adieu! Adieu! –
CHESTER. Go, generous Enemy!
And never in the field may I encounter thee!
My sword would drop, my nerveless arm forego
It's remnant vigour, while my aching Heart
At every bleeding pore, would burst to own
My Child's Deliverer, in my Country's Foe!

Exit DE BELESME.

Scene x

CHESTER. ADELA.

ADELA. My Father!
CHESTER. My Heart's joy!
ADELA. My dearest Father! –
O yet – O sir! – I trust thy noble nature –
CHESTER. What would my Child?
ADELA. He's gone – without a passport! –
CHESTER. What hoa! De Lacey! –

goes out.

ADELA. Most benign of Parents!

With what a liberal Soul has Heaven endow'd him!

CHESTER *re-enters*.

CHESTER. An Escort shall attend, and safely guard him
Back to the Citadel. – My dearest Daughter! –
Whence flow those Tears!
ADELA. Fatigue, – alarm, – long doubt, –
CHESTER. All has been sad; but now, 'twill all be joy;
Return'd, restor'd, – My Adela? – thou tremblest? –
What anxious terror mocks thus thy security
With timid fears of past, or fancied dangers?
ADELA. Blest is thy grateful Adela, – yet – oh! –
Should he who sav'd her – Martyr to his goodness –
Fall victim of Earl Mortaign's dangerous wrath –
CHESTER. How wraught he thy escape?
ADELA. Late yester-night
He came to me; told me the dreadful threat
Of Famine to the Castle; bid me prepare
Disguis'd, to follow him: then charg'd De Alderay,
His faithful Friend, to aid the enterprize,
And guard its secret. Well he knew Earl Mortaign
Would mar his generous purpose if disclos'd.
CHESTER. Yet how elude discovery?
ADELA. That he seeks not:
His dauntless nature, heedless of itself,
Means to unfold the whole, when once I am safe.
CHESTER. Perchance he trusts in th'interest of his Father,
Earl Arundel, with Mortaign?
ADELA. Should it fail! –
O pardon that I weep! – He has been so good! –
And but for him, where now were Adela? –
O not with Thee, my Father! my lov'd Father! –
CHESTER. My Bosom's Darling! what a curb is this
To rising happiness! yet it soon may pass,
And all may smile again.
ADELA. Think me not graceless –
My Heart is glad – O sir, believe it's joy!
'Tis but my spirits that droop.
CHESTER. I know it well;
Nor wonder at the cause.
ADELA. He has been so soothing,
So softening, so balsamic to my sufferings,
So feeling to my fears, – he pitied, cherish'd me,
Taught me to wile away a Captive's life.
Till the reflected comfort grew his own;

Then, when impending danger scar'd, – resign'd me.
Heaven must reward him!
CHESTER. Yes! and Heaven will!
A Parent's blessing, a fond Father's prayers
Shall follow him, shall hover o'er his Head
With guardian Gratitude, shield him from harm,
And bid him share the bliss his virtues raise.
ADELA. O words of sweetness! be they but prophetic!
CHESTER. Yet know, the Lord De Warrenne –
ADELA. O my Father! –
CHESTER. Demands ev'n now an Audience of the King
To beg thy liberty.
ADELA. May I retire?
I'm weary; – and how soft will be repose
Under thy care paternal once again!
CHESTER. May it refresh Thee! strengthen, and preserve Thee!
De Lacey!

Scene xi

CHESTER. ADELA. DE LACEY.

CHESTER. Take her to my inward Tent. –
Art thou not glad to see my Adela?
DE LACEY. Most glad in truth.
CHESTER. Go in, my love, and rest thee.

Scene xii

CHESTER *alone*.

CHESTER. Alas! De Warrenne's cause is ill assisted
By this captivity! Unhappy Times!

ACT II

Scene i

Scene the Camp. The Tent of the EARL OF CHESTER *in front.* CHESTER. FITZ HAMMON.

CHESTER. Already gone? unfortunate alacrity!
I deem'd not such facility in William.
FITZ HAMMON. 'Twas Flambard's influence; rapacious William
Had else refus'd, ev'n were the maid his own,
To save her through an offering to the foe.
But Flambard, by De Warrenne led, leads William:
For their own views, into each other's Hands
They play dominion.
CHESTER. 'Tis the fate of Tyrants:
Their vain authority is spent in words,
Mere arbitrary sounds of despotism,
While on their Creatures their effecient power,
With all its means to scourge Mankind, devolves.
FITZ HAMMON. The grace obtain'd, De Warrenne hied away;
And, more to win upon the beauteous Adela,
Carried himself his powers to the Earl Mortaign.
CHESTER. Flambard has now in charge to tell the king
Her restoration.
FITZ HAMMON. Hark! they hither come!
CHESTER. Then will I seize this moment of emotion
To plead, with joy expansive, for the Citadel;
Fair tribute to my own peculiar fortune.
FITZ HAMMON. It may be dangerous.
CHESTER. If just, no matter.
'Tis now, or never, I may speak him home.
And I will utter truths he shall not scorn
Ev'n though he may resent. My Adela!
Such be my payment for thy blest deliverance!
Exit FITZ HAMMON.

II i 12 *devolves*: three lines are deleted:
The veriest slaves that ‹keep in check› their passions
And keep them in the ‹dread› of good men's thoughts,
Controul, beguile, and turn them at their will.

Scene ii

CHESTER. WILLIAM RUFUS. FLAMBARD. ATTENDANTS *at a distance.*

WILLIAM. What do I hear, Earl Chester? Is De Warrenne
Mock'd with a useless errand to the Citadel?
CHESTER. My gracious Lord, scarce had he left my Tent
Ere Heaven vouchsaf'd me back my captive Daughter.
WILLIAM. How? and by whom conducted?
CHESTER. By a Stranger;
A generous Friend, made in her long confinement.
WILLIAM. And what her Ransom? ill can our cause endure
Largesses to the foe in times like these.
CHESTER. None have been paid: impell'd by pure compassion,
He brought her ransomless.
WILLIAM. The Tale is marvellous. –
Where is this Stranger?
CHESTER. Ere now in the Citadel.
WILLIAM. He should have been detain'd.
CHESTER. But how, my lord?
WILLIAM. By force; – in chains; – no matter how. –
CHESTER. Just Heaven! –
My Child's Deliverer! by force? in chains?
WILLIAM. When War's abroad, we list not to such forms;
Treachery is honour, then, and plunder, patriotism.
His name?
CHESTER. I meant concealment; but De Warrenne
Must now learn all. 'Tis Robert De Belesme.
WILLIAM. How? De Belesme? And have you let him go? –
The Castle's heir? – The Son of Arundel? –
And have you let him go?
CHESTER. He sav'd my Daughter!
WILLIAM. Hear then, from me, the Traitor's recompence!
He, and his Father, and each rebel follower,
With Mortaign, my base Uncle, at their head,
Within those castle Walls shall waste their being;
And die by inches under Famine's Gripe!
I'll end the struggle, teach the conquer'd people
Their vassalage; or if, at last, they yield,
Rescue nor sex nor age from general slaughter.
CHESTER (*kneeling*). O Powers benign! that in your pitying mercy
Spar'd this dread mandate to my tortur'd Ears

II ii 1 s.d. *RUFUS*: an insertion.

And lacerated Heart, till the lov'd prop
Which holds me yet on Earth, was giv'n me back,
Take Chester's thanks, in humblest adoration!
Sire (*rising*), pardon the effusion! – yet why pardon? –
O rather hear it as the cry for mercy
Piercing the Walls of yon blood-destin'd citadel.
How many bosoms there are doom'd to anguish?
How many Limbs to slaughter, Lives to massacre,
How many noble spirits to hard oppression!
How many Fathers – to bewail as I should! –
How must they feel that have not sav'd their Adelas!
That know them lost – that witness their despair –
That clasp their sinking Frames, – and wish them dead! –
WILLIAM. Hold, Earl of Chester! this is somewhat strange:
What new connections stimulate such zeal?
What is your interest, now, within those Walls?
CHESTER. The general interest of humanity!
Silent I griev'd while Adela was captive;
Now she is free, who shall arraign my motives?
O let me plead, in grateful simpathy,
For woes so nearly shar'd, – so softly 'scap'd!
WILLIAM. I hear thee with amaze! Is this fit language
For Chester? for a Warrior-Chief in arms?
To urge the cause of rebels?
CHESTER. Sir, remember
They are your subjects!
WILLIAM. Chester, they are traitors!
CHESTER. Scourge, but extirpate not! – they are your children:
WILLIAM. No, I renounce them all! – they are my Enemies!
CHESTER. Hear me, O King! – they are still your fellow-Creatures!
Man's Earthly Monarch is himself a Man;
And every subject in his proud domain
Was born his Equal, must his Equal die,
Alike expos'd at first, at last, forgotten.
The term between the Cradle and the Shroud
Marks the short limit of superior power.
How small that term, how little, how precarious!
Wouldst thou abuse such brief prerogative,
All heedless of that leveller, who strides
With hasty steps, to seize thy transient trappings,
And lay thee, all despoil'd, unfear'd, unenvied,
To crumble with the Beggar in the Dust!
WILLIAM. I stand astonish'd! whence this hardiness?
What secret spring gives birth to such bold daring?
CHESTER. A spring that issues from the source of life,
The Heart! which tells me what I had ow'd the man
Who thus had spoke, while Adela was captive.

WILLIAM. Chester, enough of Thee and of thy Adela.
The Kingdom came to me by right of Conquest;
And this disloyalty in Robert's favour
Shews me I've wielded with too mild a hand
My victor Father's sceptre: but, henceforth,
I'll make the Nation feel the Norman Yoke,
Till every rebel spirit is crush'd for-ever.
CHESTER. Hear yet one faithful Baron's voice. The king
Who seeks by terror to o'er-power his people,
Risks a wrong current to the genial feelings
Coeval with Mankind: and as they stem
Oppression's tide impetuous, the next risk
Or makes them slaves, – or ends in nameless horrours!
WILLIAM. Slaves be they, then, if such th'alternative;
Their's be the forfeit, mine, th'aggrandizement.
CHESTER. Can there be pride in trampling on the helpless?
Can there be joy in triumph o'er the vanquish'd?
Chearless the mightiest Monarch over slaves,
Mere finely form'd machines, endow'd with powers
They dare not own, with strength, they dare not use:
But void of life's inherent claim, a spirit
That conscious feels equality of soul,
Bounty divine! the birthright of humanity!
WILLIAM. Earl Chester, if wild dreams of anarchy
Amuse thy brain, at least divulge them not.
Such dangerous doctrine masks some private purpose
That spread, might instigate an insurrection.
CHESTER. Forbid it, Heaven! Tyranny and Anarchy
I hold alike abhorrent. Live the times
When Liberty and Monarchy unite!
When every sharer in his Country's concord
Shall pride, not tremble, in a just subjection.
WILLIAM. Impracticable Times! And grows this vision
From a mere casual breach in one small cage
That gives your little favourite Bird it's freedom?
CHESTER. Should I disprove the all-pervading influence
Of Home-experience, Sire, I were no man.
Yet from the vision turn not, Prince, aside –
O let me bring it nearer to your view!
He who would rule belov'd a generous Nation
Must spurn the Iron rod of despotism
To wave the Wand of gentleness and peace.

II ii 78–82 *The Kingdom ... Norman Yoke*: since William Rufus had ruled for only a few months before beginning the siege of Pevensey Castle, the mildness for which he reproaches himself was short-lived. The 'Norman Yoke' is that first imposed by his father, William the Conqueror, in subduing the Saxon people of England.

Magnetic is th'ascendance o'er mankind
Of strength with kindness blended, power, with pity.
He who would win his people's fond regard,
He who would see them follow his lov'd foot-steps,
Hear them with joy and reverence speak his name;
He who would wish his later years resplendent
With more of fame than grac'd their earliest dawn; –
Must to his virtue's solely owe his sway.
Their mild persuasion, their affecting influence
Leave far behind all arbitrary force,
And work through every fissure of the soul
To prompt allegiance by impulsive tenderness.

WILLIAM. What rhapsody is this? what wild enthusiasm
Plays on your fancy, and o'er-strains your faculties?

CHESTER. Some distant age may hail this Patriot king.
Thrice happy England, should the Hour arrive!
Blessing and blest, soft through the Vale of Time
Glide his mild course amidst his grateful people!
Where-e'er he comes, one general acclamation
Shall greet his sight, with plaudits loud and fervent,
That echo and rebound Long Live Our King!
Each beating Heart shall vibrate to the sound,
Each glistening Eye shall fill with Rapture's Tear,
Each loyal voice, with one consent, repeat
Long Live Our King, The Father Of His People!

WILLIAM. Some strange infatuation has bewitch'd Thee.
Retire awhile! –

Exit CHESTER.

Scene iii

WILLIAM. FLAMBARD.

WILLIAM. Flambard, I like not this!
I do suspect some junction with the Rebels
Has heated thus his strain'd imagination:
And, but I fear the Barons might revolt,
I would arrest him.

FLAMBARD. That, might be too violent;
'Tis through his Daughter he must be secur'd:

II ii 143 *Long Live Our King*: echoing the second line of Henry Carey's poem 'God Save the King' (c. 1745), which was rapidly adapted as the British national anthem.

II iii 5 *violent*: 'For violence too great must arouse suspicion' is deleted.

Make Adela your prize, Chester's your slave.
WILLIAM. Name me the means. His high and daring censure
Has stung me to the Soul.
FLAMBARD. My lord, demand
His instant acquiescence that De Warrenne
May wed his Daughter. In De Warrenne's hands
When once that pledge is plac'd, De Chester's fealty
Will brave all tampering from the crafty foe.
WILLIAM. So be't. And if he venture to oppose
With flighty dreams of visionary rights,
His power paternal to my kingly Will,
He shall be close confin'd.
FLAMBARD. That might inflame,
And drive the haughty Barons to Duke Robert:
Rather, my lord, send him to Rochester,
Under fair cover that his services
May there be more effectual than at Pevensey,
Where your own presence animates the siege.
WILLIAM. 'Tis better plan'd. Go, follow him thyself:
Give him th'alternative. If he will grant
His Daughter to De Warrenne, well; if not,
Command him to my Tent, for instant orders
To march himself and men to Rochester.
But tell him that, in proof of his allegiance,
I do require his Adela remain
Under my care and guidance in his absence.
FLAMBARD. From your own Tent, sir, let him then depart,
Nor see his Daughter first; for that were fraught
With dangerous combination.
WILLIAM. Ere the truce
Is o'er, I granted for De Warrenne's safety,
I have a further plan to assault the castle.
Get me a messenger.
FLAMBARD. Till De Warrenne come
Is not the truce inviolate?
WILLIAM. Ay, ay!
Or longer if they will! a second Truce!
Small favour, that allows them time to starve! –
But follow Chester; bind him to a promise,
Or bring him to my Tent for marching orders.

Exit FLAMBARD.

II iii 19 *Rochester*: Rochester Castle in Kent was also being besieged by William Rufus's forces; it fell shortly after Pevensey.

II iii 10 *De Warrenne*: 'When from his idle errand he returns' is deleted.

Scene iv

WILLIAM RUFUS *alone*.

WILLIAM. How soon were Robert's desperate cause o'erthrown
Were not these hardy Barons deep involv'd
By equal ties, with Normandy and England.
Here their acquir'd, and there, their natural rights
Keep up th'intestine interests of each country,
And mar, on either side, cordial hostility
From selfish hope to guard their double property
By double cozenage. Would the spendthrift Robert
Cede England as my due, Peace should be sign'd:
And his lax discipline, when least aware,
With scarce a blow, would give me Normandy.
De Warrenne must be pleas'd. My useful Flambard
Consults him on each mode of levying fines
On this rebellious Nation. 'Tis cheap pay
To give him Chester's Daughter for his services.

Exit.

Scene v

FITZ HAMMON. DE LACEY.

DE LACEY. Heard you the message of the King to Chester?
FITZ HAMMON. Else had I ne'er believ'd it. Surely Chester
Ill merited a mandate so suspicious.
He felt the wrong, and, stung to th'inmost soul
Hastes to the Monarch's Tent with warm remonstrance.
There, should consent be wrested by prerogative,
How ill assorted with so fair a mate
Will shew the Lord De Warrenne?
DE LACEY. O, 'tis cruel!

II iv 5 *intestine*: internal, domestic.
II iv 8 *cozenage*: deception, fraud.

II iv 10 *discipline*: the ms. reads 'disclipine'.

Scene vi

FITZ HAMMON. DE LACEY. A SOLDIER.

SOLDIER. A knight, by favour of the truce, is come,
With passport of Earl Mortaign, from the Citadel,
Seeking immediate audience of Earl Chester.
FITZ HAMMON. Bring him this way.
Exit SOLDIER.
What may his errand be?
Will they surrender? Has the Famine tam'd them
Ev'n in it's menace?

Scene vii

FITZ HAMMON. DE LACEY. DE ALDERAY.

DE ALDERAY. Where is the Earl of Chester?
FITZ HAMMON. Now with the King.
DE ALDERAY. Can I not speak with him?
FITZ HAMMON. In a few seconds.
DE ALDERAY. Sirs, delay is ruin
In business such as mine. O haste and watch
To tell the Earl, soon as he quits the King,
That on his speed to hear my embassy,
Hang Life, or Death, – with Honour, or Disgrace!
FITZ HAMMON. What rests with me, you shall not urge in vain.
Exit FITZ HAMMON.

Scene viii

DE LACEY. DE ALDERAY.

DE ALDERAY. Untimely absence! how may it involve
My injur'd friend!
DE LACEY. If Robert De Belesme
Is him you mean, I hope he not repents him
His gallant conduct to Earl Chester's Daughter?
May I enquire –
DE ALDERAY. There's nothing to conceal:

The whole is public. Lull'd by our precautions,
Suspicion slept, when, with his lovely captive,
He sallied forth, expecting to return
Ere miss'd, the bearer of his own exploit.
But suddenly a truce, by William's order,
Brought to Earl Mortaign's Tent your Lord De Warrenne.
He came for Chester's Daughter! Liberal terms
Won glad acceptance from th'impoverish'd Earl.
He summon'd De Belesme, and his fair Prize:
No De Belesme, and no fair Prize appear'd!
Eager I flew to assert his faithful loyalty –
But, in the moment, De Belesme was ta'en!
Still in disguise, just parted from his escort,
They seiz'd, and brought him Prisoner to Earl Mortaign.

DE LACEY. His gallantry deserv'd a fairer fate;
Chester will grieve –

DE ALDERAY. He must do more than grieve,
Or De Belesme will rue his services.
In vain he urg'd his innocence of Treason;
His Captive was restor'd, without a Ransom.
To instant Death Earl Mortaign had decreed him,
Had not the powerful Arundel, his Father,
Prevail'd to send me advocate to Chester
That he would clear his son of foul collusion,
By quick remittance of a splendid Ransom.

DE LACEY. How would that clear him?

DE ALDERAY. If our Treasury fill,
We may examine motives at our leisure.
De Warrenne is detain'd till my return,
Lest Rufus break the truce, and hold me Prisoner.
Can you convey these tydings to fair Adela?
And tell her that De Alderay, who oft
Bore her the passing news, while De Belesme
Engaged the foe, here pleads his cause? – perchance
She'll soothe his sufferings with some gentle message.

DE LACEY. She's but within the Tent. I'll to her straight.

Scene ix

DE ALDERAY *alone*.

DE ALDERAY. This dire delay will fill Earl Mortaign's Breast
With doubts yet deeper of foul practices:
De Warrenne will enrage at his detention;
And haughty Rufus, should he learn my errand

Ere Chester, with a bounty competent,
Dismisses me, may stay me in the Camp
Till my much injur'd friend –

Scene x

De Alderay. Adela. De Lacey.

Adela. De Alderay! –
What am I told? – speak! – De Belesme arrested? –
De Alderay. Yes! and attainted of high Treason:
And nothing less than the most liberal Ransom
For Thee, fair maid, can save his life and honour.
Adela. O I revive! – If on a Ransom rest
His life and honour, they are safe as pure!
By justice bound, and urg'd by mild benevolence,
My noble Father will delight to prove
That adverse fates may teem with kindred worth!
De Alderay. Such, too, is De Belesme's most firm persuasion:
Though Lord De Warrenne, hearing your escape,
Refus'd all terms –
Adela. O there, indeed, you glad me!
I would owe nothing to the Lord De Warrenne.

Scene xi

De Alderay. Adela. De Lacey. Fitz Hammon.

Fitz Hammon. Sorrowing I come, the messenger of ill; –
Chester is ev'n this moment gone for Rochester.
Adela. Gone? – Gracious Heaven!
De Alderay. Gone? – Then my friend is lost!
Adela. Impossible! O you are misinform'd; –
Here, in the Camp, would Chester leave his Daughter?
Fitz Hammon. 'Twas by the King's command.
Adela. The King's command?
De Alderay. Have you not told him, then, my embassy?
Fitz Hammon. Flambard forbid or conference or approach:
His troop was ready; from the Tent at once
Sudden and quick he march'd.
Adela. O then pursue him! –
Haste, good Fitz Hammon! –
Fitz Hammon. Most impossible!

A soldier's station –
DE ALDERAY. Let me then depart. –
Alas! how heavier far than Death, Dishonour
Will sting my noble Friend! –

going.

ADELA. O stay, De Alderay! –
DE ALDERAY. Have you no Jewels? – Trinkets? – Valuables? –
Nought that in lieu of money –
ADELA. Nothing! nothing! –
I live not in the Camp, nor hence was ta'en. –
FITZ HAMMON. Sir, be advis'd, and go: in duty bound
I must acquaint the King –
ADELA. O then away!
Fly! – haste to De Belesme –
DE ALDERAY. And to what purpose? –
Yet go I must, though but to seal his doom!
ADELA. O Heaven!
FITZ HAMMON. Nay, speed! or thou wilt be arrested.

Exit DE ALDERAY.

Scene xii

ADELA. DE LACEY. FITZ HAMMON.

FITZ HAMMON. And now, perforce, I must inform the King –
ADELA. Fitz Hammon stay! O hear me! aid me! save me! –
Think of some way – some means – Good, good Fitz Hammon! –
Dispatch a Courrier to my noble Father, –
Or lead me to the King, – Help me to tell him
My Honour, not my pity, is engaged! –
O quick devise –
FITZ HAMMON. Ah, madam, far too soon
The King may summon thee! –
ADELA. Why? – how? – what mean'st thou? –
FITZ HAMMON. Flambard, with studious care, beset Earl Chester,
Preventing private speech; yet, ere he march'd,
These energetic words he spoke aloud:
'I am my King's, – but Adela is mine!
'I will not force her filial gentleness;
'I go obedient, to approve my loyalty,
'But trust to find her, as I leave her, free!'
ADELA. O best of Fathers! – 'tis, then, as I thought –
Some hard compulsion of suspicious tyranny –

Scene xiii

ADELA. DE LACEY. FITZ HAMMON. BAINARD.

BAINARD. The king would see Fitz Hammon.
Exit FITZ HAMMON.
Gentle lady,
By me his highness greets Thee. State affairs
Have hurried hence Earl Chester; but he has given
His princely word to guard Thee in his absence.
Far from this scene of War he means to send thee
To safety and repose. A party arm'd
And fitted for the trust, within an Hour
Will wait thy pleasure.
ADELA. Be't as His Highness wills.
Exit BAINARD.

Scene xiv

ADELA. DE LACEY.

ADELA. Then all is o'er, and De Belesme is gone! –
O that I ne'er had quitted yonder Tower!
Ten thousand Deaths were nothing to this blow
That saves me by the loss of my preserver. –
And fall'st thou thus, thou generous De Belesme?
Diest thou for me? – accus'd! – condemn'd? – attainted? –
Is such thy fair requital? O De Lacey!
Intolerable anguish tears my Heart! –
O if thou hast any sense of Chester's honour
Save it this wound! –
DE LACEY. What, madam, can I do?
ADELA. Haste to the Citadel, – to Mortaign's feet, –
Tell him my ransom shall be sent from Rochester, –
Bid him but spare the life of my Deliverer,
And Chester will avouch his spotless fame.
DE LACEY. To Rochester I'll send; but for Earl Mortaign –
The wild attempt were Death. –
ADELA. Be it then mine! –
I'll go myself! –

II xiv 15 *for Earl Mortaign*: i.e. for De Lacey to go to Earl Mortaign.

DE LACEY. Impossible!
ADELA. Oppose not! –
'Tis all that now remains of help or hope! –
Send and acquaint my Father with my flight,
And it's occasion. – Nay, look not so forbidding! –
I shall be safe – De Warrenne still is there;
Ev'n in his hands – O Heaven! –
DE LACEY. 'Tis true, De Warrenne
Might guarantee thy safety; yet –
ADELA. Nay, pause not! –
The truce will favour me, – my old disguise
Is still at hand – aid me to pass the Guard, –
I shall be Prisoner but a little moment –
O what is danger to the call of Justice? –
My Father will approve an act of Gratitude,
And quick redeem me –
DE LACEY. But the King –
ADELA. No more! –
Only betray me not for some few minutes.
Let me elude but the return of Bainard,
Follow De Alderay, and, ere too late,
Save De Belesme! – clear his surpassing Honour, –
Then – take my chance for Ransom, – or for Death!

ACT III

A Hall within the Castle.

Scene i

EARL OF MORTAIGN. EARL OF ARUNDEL. DE BELESME. MOWBRAY. DE WARRENNE.

MORTAIGN. Earl Arundel, the War's compulsive discipline
With pain I practice on thy valiant son:
But his acquittal is I trust at hand;
De Alderay is come, and summon'd hither.
I would not see him but in this Assembly,
That thou, thy son, our Enemy, and myself
Might hear at once the fair or foul report
Which clears or blasts the fame of De Belesme.
DE BELESME. And hangs My fame on any single proof?
Have my free actions spoke so little worth,

Has my life's tenour prov'd so poor a guarantee
That Thou needst testimony to my honour?
MORTAIGN. Young Warrior yes; in times of civil feuds
I would not trust my Brother, Bayeux's Bishop,
Were he to pass the trench, as thou hast done.
ARUNDEL. But yet, Earl Mortaign, De Belesme, my son,
Saw not the tyrant Rufus; held no conference
With Flambard, the prime agent of oppression;
Simply he has deliver'd to a Father
His Captive Daughter.
MORTAIGN. Yes, – without her ransom!
ARUNDEL. 'Twas but a deed of high-wrought gallantry;
'Twas fitted to the fire of youthful heroism,
Warring for justice, yet protecting beauty,
And scoffing interest, while pursuing honour.
MORTAIGN. These sentiments, my lord, may serve to sooth
Misfortune, but they vindicate not rashness.
Unauthoris'd to quit the Citadel,
Restore a Captive of immense expectance
Unransom'd, uncondition'd, –
DE BELESME. Yet hear, Earl Mortaign!
I thought the Prize my own, my own the forfeit.
MORTAIGN. State exigence annuls all private property.
ARUNDEL. Let us then hear, my lord, Earl Chester's answer.
'Twas his own right, not thine, my Son relinquish'd;
Thy claim made known, Chester must be a Wretch
Disgracing knighthood, should he not fulfil it.
DE BELESME. Doubt not Earl Chester; the whole blame is mine.
He would have given his Earldom, nay, his life
In token of high joy. I but lament
That, paid in pleasure, I declin'd his justice,
Which now, least look'd for, I require with usury.
MORTAIGN. Thou seest the Lord De Warrenne far less scrupulous;
Impow'r'd to act, he yet avoids to treat,
Since what he meant to purchase, is purloin'd.
DE WARRENNE. Most true, and fair; 'tis to obtain we pay;
What we profess is our's on easier terms.

Enter A SOLDIER.

SOLDIER. My Lord, De Alderay is here.

III i 14 *my Brother, Bayeux's Bishop*: Odo, Bishop of Bayeux and Earl of Kent, was brother to Robert, Earl of Mortaign. Both were maternal half-brothers to William the Conqueror.

Scene ii

MORTAIGN. ARUNDEL. DE BELESME. DE WARRENNE. MOWBRAY. DE ALDERAY.

MORTAIGN. De Alderay!
The news? –
ARUNDEL. The Ransom? –
DE BELESME. Chester's answer? –
DE ALDERAY. Ah! –
MORTAIGN. How? does he hesitate? –
ARUNDEL. Has he no honour? –
DE BELESME. Deems he me, then, unworthy of redress? –
DE ALDERAY. I have not seen him; ev'n as I arriv'd
Rufus commanded him for Rochester.
MORTAIGN. Then all is o'er.
ARUNDEL. No, all is undecided.
DE BELESME. He has not, then, refus'd to succour me?
DE ALDERAY. He knew not thy arrest; and his fair Daughter
Had in the camp nor Gold, nor Jewels adequate.
DE BELESME. And thinkst thou else –
DE ALDERAY. I think she had been grateful.
MORTAIGN. No more remains; De Warrenne may depart:
But De Belesme in durance must be held
Till to Duke Robert his offence is known.
ARUNDEL. Is't possible, my lord? – Is this the meed
For his brave services? shall one youthful error
Improvident, not trait'rous, cancel all?
MORTAIGN. But for his services, and thy high influence,
Immediate Death had paid the first discovery:
At present, Bonds and Durance may suffice.
DE BELESME. Ha! Bonds and Durance? – Grant me firmness, Heaven! –
Durance and Bonds? – here, midst my fellow-Warriors? –
O cast me in some dungeon of the foe,
Consign me there to shackles, torture, Death,
Or, worse than all, endless imprisonment, –
But here – amidst my fellow Warriors – Mortaign
Disgrace me not! –
MORTAIGN. There can be no appeal
Save to Duke Robert.
ARUNDEL. Lord De Warrenne speak!
Pay down the ransom, – spare your court this shame. –
DE WARRENNE. I pay no ransom to relieve an Enemy.

III ii 13 *durance*: forced confinement, imprisonment.

ARUNDEL. Most brutal hearing! –
MORTAIGN (*To* MOWBRAY). Peace, Earl Arundel.
Take Robert De Belesme in custody,
Convey him to the Eastern Tower in bondage,
And there detain him till my further orders.
DE BELESME. Earl I obey! but by the life you leave me
You cast on me a stain ill-merited.
My Father, blush not! I will wipe it off
With my Heart's blood ere it shall blot your name.
ARUNDEL. Blush? no, my son, I glory in thy prowess,
And feel not Arundel, but the Duke suffers
When thy brave arm no more can serve his cause.
DE WARRENNE. My errand done, Earl Mortaign, I retire. –
MORTAIGN. One moment stay; and see, that thou mayst tell,
Rufus, thy Master, how his uncle Mortaign
Welcomes the Man who visits at his camp.
ARUNDEL. Injurious usage!
MORTAIGN. Arundel, be patient.
Instant in bonds put Robert De Belesme,
And take him hence – (*to* A SOLDIER – *who takes the sword of* DE BELESME.)

Scene iii

MORTAIGN. ARUNDEL. DE BELESME. MOWBRAY. DE WARRENNE. DE ALDERAY.

ADELA *rushes in.*

ADELA. No! take Earl Chester's Daughter! –
If there are bonds, 'tis her just doom to wear them:
Lo! here she stands – to claim, not deprecate them!
DE BELESME. Immortal Powers! –
DE WARRENNE. The lady Adela! –
ARUNDEL. Earl Chester's Daughter? –
MORTAIGN. What may this wonder mean?
DE BELESME. O madam! – fairest Adela! – whence com'st thou?
Can this be real? – why, and to what intent –
ADELA. Thee, thee to save – too generous De Belesme!
DE BELESME. Transcendant goodness! – At thy feet –
ADELA. No, no! –
I pay but justice, and must take no thanks.
DE WARRENNE. Has then Earl Chester sent –
ADELA. My Lord De Warrenne,
No verbal sanction ratifies my act:

'Twas Honour's impulse, quickly sensitive,
Springing from every lesson of my youth,
That taught me not to wait my Father's mandate
While his life's tenour bid me judge it's dictates.
ARUNDEL. Now, Earl of Mortaign, stands my son acquitted?
Now doom'st thou De Belesme to Bonds and Durance?
MORTAIGN. No, he's released.

(*The* SOLDIER *gives* DE BELESME *his sword.*)

ADELA. My duty's then fulfill'd.
DE BELESME. O matchless excellence! My life I priz'd not
Oppos'd to thy fair safety; – but my Honour –
For that, O madam, deign t'accept my homage!
ADELA. Suffer me now, Earl Mortaign, to implore
Your high protection, till my noble Father,
Hearing my enterprize, redeem my freedom.
MORTAIGN. Every respect due to thy birth is thine.
DE WARRENNE. Nay, madam, never seek protection here:
The powers with which I came, I now will use:
Earl, since again this lady is thy captive,
I do acede at once to pay her ransom,
So I may be her convey to the Camp.
MORTAIGN. My Lord De Warrenne, thou art now too late;
Twice captive, twice her ransom must be paid.
DE WARRENNE. Earl Mortaign, this rapacity –
MORTAIGN. No contest!
Deserve, or quit the Prize.
DE WARRENNE. I would agree
To any payment for her fair possession;
But here I have nor powers, nor means, to settle
Demands of such exorbitance.
MORTAIGN. Enough;
Thee I detain not; – but here rests the lady.
DE WARRENNE. Nay, nay, I cannot leave her; wilt thou forward
A special messenger to William's Tent?
One of my escort, whom he can entrust
With these thy hard conditions?
MORTAIGN. Be it so;
And when, prepar'd for treaty, he returns
Ye shall depart together, and in safety.
Till then, ye must abide in separate custody.
Mowbray, take to thy charge the Lord De Warrenne.
DE WARRENNE. Fair Adela, I hope to free thee soon.
Exit DE WARRENNE *with* MOWBRAY.

Scene iv

MORTAIGN. ARUNDEL. DE BELESME. DE ALDERAY. ADELA.

MORTAIGN. To thee, Earl Arundel, I next consign
The Lady Adela.
ARUNDEL. To me, my lord?
MORTAIGN. Yes, Earl; receive this test of confidence,
And banish acrimony and resentment.
ARUNDEL. Earl Mortaign, thou hast found the means to sooth
My troubled Honour. I accept the pledge.
Madam, with me remain in full security
Till Lord De Warrenne be impower'd to claim thee.
Exit ADELA *with* EARL ARUNDEL.

Scene v

MORTAIGN. DE BELESME. DE ALDERAY.

DE BELESME. Earl Mortaign, by this gallantry of trust
Thou hast won resentment from my soften'd mind. –
MORTAIGN. Most valiant De Belesme, pardon the trial.
Suspicion of thy spotless fealty
In Robert's cause, ne'er reach'd my inmost thoughts;
'Twas but a stratagem, devis'd t'extort
Some treasure from the foe. Dreadful our state!
And should no quick relief – Ha! – (*a shout without.*)

Scene vi

MORTAIGN. DE BELESME. DE ALDERAY. MOWBRAY.

MOWBRAY. To arms, Earl Mortaign! –
Treason is at the Gates. I had but station'd
De Warrenne with his Guard, when a loud shout
Announc'd confusion. –
DE BELESME. Where? – which way? –
MOWBRAY. To the West. –
Just where Earl Arundel, with his fair charge
Are crossing to his Tower.
DE BELESME. Earl Mortaign, now
I'll earn thy fuller faith! –
Exit with his sword drawn.

Scene vii

MORTAIGN. MOWBRAY. DE ALDERAY.

MORTAIGN (*to* MOWBRAY). Here keep thy watch,
While I go learn to what this tumult tends.
Exit MORTAIGN *and* DE ALDERAY.

Scene viii

MOWBRAY *alone*.

MOWBRAY. Old as I am, I like not this tame work.
Would I were with the thickest of the foe!

Scene ix

MOWBRAY. ARUNDEL. ADELA.

ARUNDEL. Mowbray, to thee I give awhile this lady.
There is no passing yet for wild commotion.
Madam, take no alarm; 'twill soon be quell'd;
And for thy noble conduct to my son
I shall regard thy safety as my own.
Exit ARUNDEL.

Scene x

MOWBRAY. ADELA.

MOWBRAY. These are ill-sorted times for gentle ladies;
I marvel, madam, thou couldst come abroad.
Though Rufus, whom thou callst thy king, might drive
Ev'n Babies from their mother's Laps, to seek
With Mortaign shelter, in Duke Robert's cause.
ADELA. O that these dire, unnatural feuds were o'er!
Conquest has here the horrours of defeat;
On either side we view alike our friends,
And dread a victory while we court success!

MOWBRAY. The fault is all in Rufus; by his crafts
He gain'd the conqueror from his eldest son;
For what has he to do on England's throne?
'Tis Robert's right, – if it belong to either; –
But in good faith 'tis Edgar Atheling's,
For Birth holds right, though conquest gives possession. –
DE BELESME (*without*). Where – which way is the Earl of Chester's Daughter?

Scene xi

MOWBRAY. ADELA. DE BELESME.

DE BELESME. Still, madam, here? – Good Mowbray, I entreat
Haste to the Ramparts; – in my Father's right
Myself will guard this lady, and convey her,
Soon as the crowd disperses, to his Tower.
MOWBRAY. But whence was the alarm? –
DE BELESME. A spy, disguis'd,
From Rufus' camp, was suddenly detected.
MOWBRAY. Was that the tumult? – for a single man? –
I thought a Regiment at least at hand.
DE BELESME. So thought the men on guard; and their suspicion
Caus'd the commotion. But go hear the rest
With Mortaign and my Father, who, in council,
Are hastening to examine him.
MOWBRAY. Methinks
Thou art wond'rous eager to dismiss me hence:
But as I once was young myself –
DE BELESME. Good Mowbray,
Why wilt thou thus lose time?
MOWBRAY. Well, well, I'm gone. –
Exit.

III x 14–15 *Edgar Atheling's ... possession*: Edgar Atheling's claim to the throne as successor to Edward the Confessor was lost with the accession of Harold in 1066. A brief attempt by Edgar to assume power after Harold's death at the battle of Hastings was also unsuccessful. The heroine of *Elberta* is the daughter of Edgar Atheling's variously named second-in-command, Ethelbert or Elwolf.

Scene xii

ADELA. DE BELESME.

DE BELESME. O madam! pardon if with haste abrupt
I utter words that with severest caution,
All fearful to offend, I have, thus long
Rigourously withheld! – denied myself the solace
To speak ev'n of my sorrows, or my wishes, –
Forbore but now, when my whole bounding Soul
Sprung forward, to pay tribute to thy virtues,
Forbore, – in dread respect – to speak – to utter –
ADELA. Still, still forbear! –
DE BELESME. Ah madam, breathe not now
An interdiction fatal to my peace! –
A few short Hours may part us ever more! –
The spy but just detected, brought dispatches
From Flambard to De Warrenne, with instructions
Most treacherous 'gainst the citadel; concluding
With this bold promise, made in Rufus' name:
'Chester is gone, – and Adela's thy own.'
ADELA. O Heaven! –
DE BELESME. Ah speak – lovst thou the Lord De Warrenne?
ADELA. Love him? – I sicken at his very name! –
Love him? – I shrink with horrour at his sight!
DE BELESME. Then, Adela, – O fairest, loveliest! – hear me!
I do not beg thee to forswear thy sovereign;
I do not ask thee to forego thy principles;
I do not wish thee to desert thy Father –
ADELA. O De Belesme! – I dread to hear thy purpose! –
DE BELESME. I but solicit this; – preserve, defend
Thy Hand from vows unhallow'd by thy Heart!
ADELA. Yes, there I pledge my faith! My noble Father
Was never form'd to link with Lord De Warrenne.
Far be our hands, as are their souls asunder!
DE BELESME. Thou yet perceiv'st not where the peril lies:
De Warrenne takes thee hence, not to thy Father, –
No, – to the tyrant Rufus in his absence:
Who, then, and what, may save thee from their Will?
ADELA. My firm denial.
DE BELESME. Ah, sweet Adela,
Canst thou resist the various arts of power
If through thy Father's safety they assault thee?
ADELA. Dreadful idea! present not such a conflict
To my affrighted thoughts!

DE BELESME. Ah! – thou wouldst yield!
ADELA. Thou couldst not wish me to forsake my Father?
DE BELESME. And can I wish thee Wife of Lord De Warrenne?
ADELA. O hateful sound! O never rise the Sun
On that black Hour which falsifies my vows!
DE BELESME. One way alone remains –
ADELA. Speak it then quick!
DE BELESME. O Adela! – and wilt thou deign to hear it? –
ADELA. In fear, alas! – yet, what wouldst thou propose?
DE BELESME. Be mine, sweet maid, at once, – and all is safe.
ADELA. O De Belesme! – and springs this plan from thee? –
From thee – in whom the lustre of fair Honour
I thought enshrin'd in pristine purity?
Superior to each guile of self –
DE BELESME. Hold, Adela!
Deem me not modell'd in a mould so sordid
That self alone prompts my presumptuous wish: –
Hast thou not own'd thou dost not love De Warrenne?
ADELA. Love him? – the Grave is dearer to my thoughts!
DE BELESME. Can I endure to see thy life, thy beauty,
The victims of compulsion? Thy mind's choice
Driven from it's rights? Thy Time confiscated
By heartless duties? Thy fair Will the property
Of Avarice and baseness? Of a Wretch
Who sees thy Beauty, and who knows thy dowry,
And leaves unquestion'd all the kind accordance
Of mutual tastes, and grateful simpathies? –
ADELA. Alas, he is stranger even to their meaning.
DE BELESME. Could I behold thee thus the prey of tyranny,
Ev'n though rejection kill'd me from thy Eyes,
And told me, bound or free, thou wert alike
To me denied, inflexible, and cold?
No! this right arm must lose the power to act
In concord with my will, ere I can wait
The glozing of Reward, to aid distress.
ADELA. How couldst thou, then, with principles thus just,
With feelings thus refin'd, How couldst thou name
A deed clandestine? –

III xii 70 *glozing*: flattery, speciousness.

Scene xiii

ADELA. DE BELESME. MOWBRAY.

MOWBRAY. And is this thy haste?
Methinks, my Warrior, when thou sentst me forth
Thy humour seem'd of greater expedition.
The young, howe'er, are volatile. Fair lady,
Thou hast much to learn of such vicissitudes.
The stoutest chieftain may forget his calling
To wait the wily leisure of thy smiles:
But, trust me, when he has found the road to win them,
His memory will find her's for other matters.
DE BELESME. Good Mowbray, werefore hast thou left the council?
MOWBRAY. Good De Belesme, wouldst have the Council last
Till Time hang heavy on thy hands this way?
By such a reckoning, we might chance to starve;
And sure these times have other means of Death
To satisfy our cravings for the Grave.
DE BELESME. Where is the spy? what more have they discover'd?
MOWBRAY. The spy has paid his Ransom with his life;
For having within reach nought else to offer
Earl Mortaign took it as the readiest fine.
DE BELESME. Did nothing more transpire?
MOWBRAY. Some small confession.
DE BELESME. And to what purpose?
MOWBRAY. Touching this fair captive.
ADELA. Me? –
DE BELESME. O declare it!
MOWBRAY. On my life, young soldier,
Thou longst to make me raise the lady's blushes. –
ADELA. Explain, I beg!
MOWBRAY. Nay, madam, 'twas but news
Importing Chester's absence made all smooth
To combat with each maiden obstacle,
And frame De Warrenne for a happy Bridegroom.
ADELA. O Heaven! O gracious Powers! –
DE BELESME. Fly, Mowbray, fly, –
See if the Coast is clear –
MOWBRAY. What! fall'n again
Into thy mood for speed? said I not, lady,
Thou hadst much of such vicissitudes to learn?
DE BELESME. Nay, prithee run, good Mowbray! –
MOWBRAY. Run? Sir knight?
I'm glad I look so nimble! On my life

I knew not I had such an airy semblance.
Run? –
DE BELESME. Trifle not, I pray!
MOWBRAY. Well, well, I'll go:
Ere I return, thy speed may slack again.

Scene xiv

ADELA. DE BELESME.

DE BELESME. O Adela! – o yet – wilt thou not hear me? –
ADELA. Alas! Alas!
DE BELESME. Thou seest the instant danger –
ADELA. O every way undone! – o my lov'd Father! –
DE BELESME. Still thou art his. I pledge to thee mine Honour
I ne'er will claim thee from the noble Chester.
Give me but right to whisper to De Warrenne
He takes not to the Camp his destin'd Bride.
ADELA. His Bride? – o never, never! –
DE BELESME. Bayeux' Bishop
Is my fast friend: might he but plight our faiths
With holy benediction! –
ADELA. O my Father! –
DE BELESME. Hadst thou but seen him, when with trembling awe,
Distant, I ventur'd to suggest a wish –
Thy fears would die away; – his manly cheek
Glow'd, with no angry flush, – his open front
Benign concurrence spoke –
ADELA. Tenderest of Parents! –
Can I, without thy sanction –
DE BELESME. View th'alternative! –
De Warrenne –
ADELA. O thou seest my Heart abhors him! –
DE BELESME. Rufus must know thee mine – or make thee his; –
And wilt thou, then, –
ADELA. O dire and dreadful crisis! –
DE BELESME. Deign, deign, O Adela! to hear, and trust me.
Though I would War with myriads to obtain thee,
If there is nothing pleads for me but prayer –
I've done! O let me not oppress thy softness! –
If no esteem, no kindness of the Heart
Point out, untold, that bliss diffus'd inspires
Refin'd delight, ev'n in the Breast fastidious
That tastes but by reflection human joys –
Speak! – pardon my presumption? – and dismiss me

To anguish keen and deep – yet to submission.
ADELA. O De Belesme, my Heart is rent asunder! –
Thou hast a Father! –
DE BELESME. Yes, sweet Adela,
And well he knows the ambition of his Son. –
ADELA. But such a Father! – holds the World his Equal?
Duty with me stands second towards obedience;
Affection, urg'd by reverence of his worth,
Takes the first place, and so impels compliance,
That, while my wishes blend with his commands,
I scarce know which direct!
DE BELESME. Thou still art his!
Have I not vow'd –

Scene xv

ADELA. DE BELESME. MOWBRAY.

MOWBRAY. Knowst thou Earl Arundel
Already in his Tower, waits thy coming?
DE BELESME. So soon? May I, then, madam, lead thee thither?
Exeunt DE BELESME *and* ADELA.

Scene xvi

MOWBRAY *alone*.

MOWBRAY. So! he again has chang'd his boisterous hurry
For sluggish sloth! – He is not much in fault:
The time has been I might have done the same.

ACT IV

Scene i

Scene An Apartment in the Tower of the EARL OF ARUNDEL.

ADELA.

ADELA. O dreaded act! – Will Heaven such nuptials bless? –
The first great Law of life, filial obedience,
Broken – revers'd! – How will he bear to hear it? –
His Child – his age's promise – Thus unsanction'd
To change parental for connubial bonds! –
And with an Enemy? – O De Belesme
Why do I mix thy kindred simpathies
With enmities of state? Were but the times
Propitious to thy virtue, the wide universe
Holds not a son so fitted to De Chester;
So like in Honour, so complete in courage,
Of spirit so lofty, yet of Heart so gentle!

Scene ii

ADELA. DE ALDERAY.

DE ALDERAY. Madam, I greet thee from Earl Arundel.
De Warrenne's messenger has brought from Rufus
The double ransom. Mortaign is content:
And, with the Morning's sun De Warrenne claims thee.
ADELA. I must, perforce, be ready to attend him.
DE ALDERAY. A softer message next entreats thy Ear:
Soon as the shades of night involve the skies
Hither, with thy fair leave, I shall conduct
Good Bayeux' Bishop, with my gallant friend,
To whom I hope, in semblance of a Father,
To give a treasure –
ADELA. Utter not that name! –
O name rever'd! – or all my courage fails!

IV ii 7 *involve*: envelop, enfold.

DE ALDERAY. Canst thou believe Earl Chester then repugnant
To valiant De Belesme?
ADELA. Could I think that –
De Warrenne's self would make me less unhappy! –
DE ALDERAY. What vain alarm –
ADELA. Not vain! – To take a step
Never to be recall'd, unauthoris'd
By his consent – unhallow'd by his blessing! –
How may I meet his Eye? –
DE ALDERAY. Grant De Belesme
The task of explanation; he conjures it,
And craves thy Word none other than himself
May send, or tell, the tydings.
ADELA. Grateful I feel
His generous wish to spare my conscious fears;
Nor will I thwart it: be it as he will;
I leave it to him wholly; take my promise.
Alas! how could I speak myself?
DE ALDERAY. Think, lady,
'Tis thy sole refuge from tyrannic force.
Still from this scene of famine and despair
De Warrenne may convey thee, when fast vows
Have given to De Belesme –

Scene iii

ADELA. DE ALDERAY. MOWBRAY.

MOWBRAY. The Lord De Warrenne
Begs license that his messenger from Rufus
May speak with thee unwitness'd. I have told him
Many a messenger might wish the same.
ADELA. His embassies to me are all unwelcome.
DE ALDERAY (*apart to* ADELA). When midnight tolls, the Bishop and my Friend –
MOWBRAY. Why dost thou loiter? prithee speed thy march.
Come! or I'll send thee Robert De Belesme;
He'll haste thy steps, – – howe'er he slack his own.
Exeunt MOWBRAY *and* DE ALDERAY.

Scene iv

ADELA *alone*.

ADELA. And now 'tis past recall! O let me hope
Not past forgiveness! – This importunity
Augments my fixt aversion to De Warrenne.
O that my generous Father knew my conflict!
Would he not say Take De Belesme – and end it!

Scene v

ADELA. CHESTER, *disguised*. MOWBRAY *conducts him in, and retires*.

ADELA. Thou comst from Lord De Warrenne? – Speak his pleasure.
– Art thou but now from the king's camp return'd? –
I fear thou bringst me thence some news unwelcome? –
Say! – knowst thou aught of Chester's Earl?
CHESTER. My Adela! –
ADELA. O Heaven! –
CHESTER. My child! –
ADELA. O powers divine! – I die! –
CHESTER. Be not alarm'd; I come but to protect thee;
Unknown to all – why art thou thus dismay'd? –
Revive, my child, revive! –
ADELA. O hide me! hide me! –
CHESTER. Compose thy spirits, or we're both betray'd.
De Lacey's tydings, by Express, o'er-took me
While on my march towards Rochester. To hear
That thou again wert captive, tore my Heart! –
ADELA. Oh! –
CHESTER. Instant I delegated my authority
To an approv'd brave officer; then turn'd
My hasty steps this way, – where pitying Fortune
Directed me to meet De Warrenne's messenger.
I borrow'd from him this accoutrement
With which in safety I assum'd his office,
And to De Warrenne sped with feign'd dispatches.
Hither, through his connivance, am I come,
And our joint care conducts thee hence i' the morn. –
Art thou not well? – why dost thou drop thy Eyes? –
Why art thou silent? – why thus pale? – thus chill? –
ADELA. O thou canst ne'er forgive – ah! look not at me! –

CHESTER. Dismiss thy fears; – thy flight indeed was rash,
Ill-tim'd, and perilous; yet not dishonourable:
I felt it's impulse with accordant strings,
And, feeling, pardon'd it.
ADELA. O speak not thus! –
This goodness stabs me! –
CHESTER. No, be comforted;
'Tis wholly past.
ADELA. Alas! – thou knowst not all! –
CHESTER. Not all? – what dost thou mean? – what all? –
ADELA. Oh – ask not! –
Or ask not now! –
CHESTER. Nay, tell the whole at once.
Concealment is in youth the nurse of danger;
'Tis Art's first step, simplicity's first fall:
Where early practic'd, 'tis at virtue's peril;
Candour is sullied; Innocence foregoes
That delicate, yet dauntless openness
Which rather does not know, than shuns disguise.
ADELA. O I would fain speak all – but fear to involve –
Alas! alas! –
CHESTER. I know thee, Adela,
To evil all a stranger, to deceit
Invulnerable; yet see I some deep purpose
Framing, or fram'd, perform'd, or to perform –
ADELA. Ah, press not now – Earl Arundel may marvel
Thou stayst so long. –
CHESTER. Dost thou then wish me gone?
ADELA. O no! – and yet – O could I send a messenger –
CHESTER. Why wilt not open to me all thy heart?
Fear not to speak; thou shalt not find me rigid:
I did not rear my child to make her wretched;
When Nature gave me power, it gave me tenderness
To guard it from abuse.
ADELA. Heavenly benignity!
O that I now could cast me at thy feet,
Trust thee with every thought – expose each wish –
CHESTER. What holds thee back?
ADELA. Alas – I stand not singly, –
And should I injure –
CHESTER. Whom?
ADELA. – I cannot speak! –
O couldst thou grant me yet a few short moments
To gain my Word's release – to clear my conscience,
I might be then more blameless in thy sight,
And with recover'd courage bare my Breast. –
CHESTER. Do as thou wilt. I aim not at compulsion.

Thy Honour is the Honour of my House,
As such, I mean to guard it with my own:
But for thy confidence – there, Adela,
I leave thee to thyself.
ADELA. But not in anger?
Say not in anger! spare my struggling soul! –
O dear to me as are the springs of life –
And dearer far, for life has lost its charms
With innocence, –
CHESTER. How! Art thou guilty? – Thou! –
ADELA. I know not what I am! Duty, Affection,
Honour, and Gratitude –
CHESTER. Are they at War? –
Who, what has sunder'd them?
ADELA. O! I am bound –
Rash, fatal promise! –
CHESTER. What is it thou hast done?
ADELA. How shall I tell! – O curse me not, my Father! –
CHESTER. Curse Thee? O Adela! not if thou murder'st me!
Curse thee? Thou shakst my soul with wondering horror! –
What hast thou done to fear my curse?
ADELA. Ah! – spare me! –
CHESTER. I will! I know thy honour, and know well
Some sophistry must paint it still preserv'd
Ev'n as thou cancel'st it: yet, thou'rt misled:
What Honour joins duplicity with Duty?
Or vindicates its breach, if no injustice
Dissolve the ties of primitive protection
By the redeeming Law of general Right? –
ADELA. Deign but to wait till I call back my promise –
CHESTER. Hear yet a Word. If in the secret purpose
With which thy bosom trembling teems, thy happiness
Appear the end, or seem to thee envellopp'd –
Beware, in time, beware, of self-delusion! –
I know thou dost not mean to lose thy Father!
ADELA. O Heav'n! –
CHESTER. Those who would win thee to unfilial deeds
May plead thy pow'r o'er my parental feelings
To ensure forgiveness; and conscious favour
May whisper in thy Ear thy Father's fondness
Could not behold thee suppliant and in anguish
Yet fiercely turn away inexorable.
Be this presum'd: I seek not to disclaim it.
Yet hear me, thou poor Adela! –
ADELA. O Sir! –
CHESTER. Suppose thy full success, suppose thee pardon'd –
Say, is it nothing to require forgiveness? –

ADELA. O gracious Powers! – o that the whole were known! –
CHESTER. Nothing, to lose the innocent elation
That, bounding in thy youthful Breast, till now,
With virtue's rapture, juvenile and gay,
Wak'd thee each morn with this fair filial Hymn,
And with it clos'd each Night:
'My Father's Love is mine, his love deserv'd,
For I have ne'er offended him. – '
ADELA. Great Heaven! –
CHESTER. 'I blest his latter Days, I sooth'd his cares,
He call'd me all his joy – '
ADELA. O cease! o cease!
O my lov'd Father! rend not thus my Heart!
CHESTER. No, I have done. I see some general evil,
And know the grateful ties that bind thee fast.
I leave thee to thyself, – with one short prayer, –
That Heaven, O Adela, which form'd thee pure,
Which gave thee to my fondness all it wish'd,
And lent me thy first youth in perfect innocence,
May guard thee still, may pity thy distress,
With strengthen'd virtue terminate thy conflict,
And spare this aching Heart to beat one moment
Beyond the term that sees thee all unsullied,
All angel-white! – all blessing – where thou art blest –
The anxious parent who now folds thee fast
To his supporting Bosom! –
ADELA. There let me rest!
O take me all! take me to thy soft mercy! –
The conflict's o'er, – I yield me to thy guidance –
Hear, and forgive thy erring Adela!
O let me cling to Thee till I regain
My lost felicity! my single duties!
My undivided ties! my early confidence!
Primaeval blessings! born in tenderness,
Nourish'd in filial joy, unstain'd, unmixt, –

Scene vi

ADELA. CHESTER. DE BELESME.

DE BELESME. Earl Chester! –
ADELA. Ah! –
CHESTER. What means this? – De Belesme? –
Am I then known?
DE BELESME. Fear nothing. –

ADELA. Gracious Heaven! –
DE BELESME. O start not thus! I come not to distress Thee;
No, beauteous Adela, I see, and venerate
The high protection that redeems my claim!
I come to rescue and unite, not sever you! –
My lord, You must away! the base De Warrenne
Has treacherously betray'd you to Earl Mortaign,
Who in an instant will pursue you hither.
CHESTER. De Warrenne!
ADELA. O inhuman! –
DE BELESME. His intent
Is, doubtless, to detain you by arrest
Till of his prize secure. From De Alderay,
Present himself, I heard the perfidy; –
And flew to save you, ere my Father learn
Your name and rank. Haste from the citadel –
CHESTER. And leave my Daughter? –
ADELA. O my lov'd Father! –
DE BELESME. No!
Take her, my lord! In your parental care
Her peace and purity alone are safe.
CHESTER. Are you so generous?
DE BELESME. Rather say, so just.
All, here, is peril – can I wish her stay?
You will not pay a vilain with her hand –
I ask no more! –
CHESTER. Honour can go no further.
My sense of Duty to my king and country
Forbid all secret ties.
DE BELESME. My lot severe
Shall not arraign your justice. Haste, my lord, –
Danger, disgrace, and Death attend delay!
De Alderay below prepares your route:
He knows a private passage that may lead you
Safe on the road to Rochester.
ADELA. Farewell! –
DE BELESME. You would be lost this perilous way alone:
I'll guide you near the Ramparts, to De Alderay;
Thence, as you fly, I may descry your safety:
'Twill sooth me as we part! –
CHESTER. Thrice noble youth!
Thanks were too poor, – be thine the mede of Honour
Which teaches Virtue to defy calamity!

Exeunt all three by a private Door.

Scene vii

Enter ARUNDEL.

ARUNDEL. Madam – Not here! – Gone! – And the Messenger! –
Astonishing! – Both gone! – Mowbray! what hoa! –

Scene viii

ARUNDEL. MOWBRAY.

ARUNDEL. Where is the lady Adela? – And where
The Lord De Warrenne's messenger?
MOWBRAY. Most strange! –
Are they not here?
ARUNDEL. Canst thou not see?
MOWBRAY. Most wonderful!
'Tis scarce a minute that I left them both!
ARUNDEL. A minute? no; his length of stay alarm'd me –
MOWBRAY. To say the truth, I thought that messenger
Had somewhat rather covert in his look. –
ARUNDEL. Call me my son. – I pray thee, Mowbray, speed!
MOWBRAY. Nay, good my lord, begin not with your speed!
Leave that to De Belesme; he'll take good care
I shall not make the night grow old with sloth.

Exit MOWBRAY.

Scene ix

ARUNDEL *alone*.

ARUNDEL. Incomprehensible! the messenger
Could teach no private avenues from the Tower;
Stranger himself – Who's that? –

Scene x

ARUNDEL. MORTAIGN.

MORTAIGN. Earl Arundel,
Where is thy captive? – where the messenger?
ARUNDEL. With deep dismay, I miss them both this moment. –

MORTAIGN. Quick let them be pursu'd! – Call me De Warrenne! –
Send De Belesme himself to seek the fugitives! –

Exit ARUNDEL.

Scene xi

MORTAIGN *alone*.

MORTAIGN. Treason is rife each way! I see our fall!
I see my injur'd Nephew's overthrow!

Scene xii

MORTAIGN. MOWBRAY.

MOWBRAY. Nor quick, nor slow can De Belesme be found.
But here's the Lord De Warrenne.
MORTAIGN. How? not found? –

Scene xiii

MORTAIGN. MOWBRAY. DE WARRENNE.

DE WARRENNE. My lord I come, by virtue of her ransom
To claim the lady Adela.
MOWBRAY. Then take her! –
If thou canst find her, I believe thou'rt welcome.
MORTAIGN. Fresh treachery every moment opens round me! –
Why have I urg'd Duke Robert to resistance? –
Call back Earl Arundel! – send for De Alderay! –

Exit MOWBRAY.

What evil is not short of civil War?
DE WARRENNE. Has she again escap'd?
MORTAIGN. Nay, so it seems; –
If so it also prove, and if this messenger,
This Earl disguis'd, has thus eluded us,
We must prolong thy stay a further term.
DE WARRENNE. How so, my lord? I am not here a Captive;
I come but on a peaceful embassy.
MOWBRAY. No matter; if the Earl is gone, 'tis fitting
One of his party make us reparation.

Scene xiv

MORTAIGN. DE WARRENNE. ARUNDEL.

ARUNDEL. Confus'd, amaz'd, I stand again before thee:
My son has disappear'd! – no one knows how,
No one can tell me of his route, or purpose!
MORTAIGN. Nay, then, – if De Belesme becomes a Traitor
Where may Duke Robert look for truth and loyalty!

Scene xv

MORTAIGN. DE WARRENNE. ARUNDEL. MOWBRAY. DE ALDERAY.

MOWBRAY. My lord, I bring De Alderay as a Prisoner.
MORTAIGN. De Alderay? –
ARUNDEL. The friend of De Belesme? –
MOWBRAY. Ay, and the would-be partner of his flights,
But that, detecting him on Ground suspicious,
I brought him hither from his wild associate
To company more sober.
MORTAIGN. Speak, De Alderay,
Knowst thou of De Belesme?
ARUNDEL. Where is my son?
DE ALDERAY. Where-e'er he is, of this be sure; his Soul
Is spotless from black treachery; his pursuit
Is Honour; and though dire mischance obscure
The brilliant prospects of his sanguine views,
No cloud will taint the colour of his mind.
ARUNDEL. My valliant son! –
DE ALDERAY. His life he holds at nought
When Friendship or Distress exertion claim:
Ev'n now, he falls the victim of high heroism –
ARUNDEL. He falls? – My Son? –
MORTAIGN. De Alderay, explain!
DE ALDERAY. Not in the toils of Death, but of captivity.
ARUNDEL. Oh cruel Destiny! –
DE WARRENNE. Follow'd he Adela?
MORTAIGN. Pursu'd he the false messenger, Earl Chester?
DE ALDERAY. Both were already flown beyond recall
When from the Ramparts we a tumult heard:
A piercing shriek was ecchoed to our Ears:
It seem'd of Adela, that moment ta'en; –
Down darted De Belesme – he thought the seizure

Made by our Guards, to hinder the escape:
Fearless and unsuspicious, on he rush'd, –
The sable Night concealing his mistake;
But, as I eager follow'd to his rescue,
I found a party of the Tyrant Rufus
In ambush lying, had assail'd my friend.
And, with the Earl and lady, made him captive!
ARUNDEL. O most unhappy! –
MORTAIGN. 'Twas a wilful error!
Bent to protect the fugitive, he flew
Madly precipitate, to certain thraldom. –
DE WARRENNE. May I not now, Earl Mortaign, –
MORTAIGN. Mowbray, hence
Remove the Lord De Warrenne to safe keeping.
DE WARRENNE. This is a deed –
MOWBRAY. Nay, good my lord, no parleying!
Exeunt DE WARRENNE *and* MOWBRAY.

Scene xvi

MORTAIGN. ARUNDEL. DE ALDERAY.

ARUNDEL. O unexpected chance! My son! My son! –
MORTAIGN. General destruction seems suspended o'er us,
And threatens quick an overwhelming fall! –
Why did I ever counsel civil War?
O desperate evil! keenest of endurance,
Yet of deliv'rance hardest! fraught with woe,
Desertion, Famine, dire distress, and Treason!

ACT V

Scene i

Scene, the Tent of the King.

WILLIAM RUFUS. FLAMBARD.

WILLIAM. For this defection stood I least prepar'd:
Chester, with all his hardiness of speech,
Seem'd still of loyal mould. Gone to the Foe!

V i 3 *Gone to the Foe!*: an insertion.

FLAMBARD. Yet the Deserter,
Sent by De Warrenne with this special notice,
Could not mistake the well known name of Chester; –
The account, in all its parts, assures veracity.
WILLIAM. When, with yon famish'd tribe, which soon must yield,
He falls again beneath my lawful sway,
How will I glut and satiate my revenge!
He shall not live an Hour. His Daughter, too, –
What punishment befits her daring flight?
FLAMBARD. She'll find a Tamer in the Lord De Warrenne.
What most disturbs me, is the double ransom
Exacted by Earl Mortaign. –
WILLIAM. 'Tis a fraud
That but to serve De Warrenne, I had laugh'd at. –

Scene ii

WILLIAM. FLAMBARD. BAINARD.

BAINARD. My Leige, Earl Chester and the lady Adela
This moment are secur'd.
WILLIAM. Most welcome hearing!
BAINARD. Nor is this all; a Warrior, from the Ramparts
Rushing to save them, fell into our hands,
And proves the potent Robert De Belesme.
WILLIAM. Thou givst me joy on joy. I'll rid myself
Of both at once, Chester and De Belesme.
Let the same fate cement their upstart friendship.
Go! see that, hand in hand, they die together
Facing the Citadel, whence the starv'd Garrison
May view, and envy their auspicious exit.

Exit BAINARD.

Scene iii

WILLIAM. FLAMBARD.

WILLIAM. This is beyond my hopes. Chester attainted,
His Earldom and Estates fall to the Crown:
And De Belesme destroy'd, each Rebel trembles.

Scene iv

WILLIAM. FLAMBARD. DE WARRENNE.

WILLIAM. De Warrenne? And alone? How is't thou'rt come,
With double ransom paid, and yet no Damsel?
DE WARRENNE. My Liege, the marvel is I come at all.
Chester and Adela, by means unknown,
With Robert De Belesme, last night absconded.
Report proclaim'd them taken by your Troops.
Mortaign, with needy gripe of further treasure,
Detain'd me Prisoner: but when he and Mowbray
Retir'd, my famish'd Guard could ill resist
A proffer'd bribe to aid in my escape.
Are not the fugitives in custody?
WILLIAM. They are. What thinkst thou now of Chester's Daughter?
DE WARRENNE. That she is fair, and rich.
WILLIAM. And does her scorn
Nothing abate thy passion?
DE WARRENNE. Scorn, my Liege,
Is but the froward pride of early beauty,
Soon set apart by those who know to govern.
WILLIAM. But should her Father's fortune be confiscate?
DE WARRENNE. My lord, I hold you to your royal promise
That she and her full dowry shall be mine.
WILLIAM. I knew not then that Chester was a Traitor.
DE WARRENNE. Nay, good my lord, if you retract – I've done!
walks away.

FLAMBARD. Pardon, my liege, but hear me! –
WILLIAM. Speak!
FLAMBARD. De Warrenne,
In times thus critical, if made a foe –
WILLIAM. What! am I menac'd? –
FLAMBARD. Nay, my lord, remember
Much is his due for services unpaid. –
WILLIAM. Well, – keep him quiet; – follow, and appease him.
Exeunt FLAMBARD *and* DE WARRENNE.

V iv 15 *froward*: perverse, refractory.

Scene v

WILLIAM *alone*.

WILLIAM. Had I foreseen this splendid confiscation
De Warrenne might have sought another Bride.

Scene vi

WILLIAM. BAINARD.

BAINARD. My liege, a dangerous tumult seems at hand.
All is now ready to fulfil your orders,
But every soldier, rising for Earl Chester,
Threatens revolt should he be executed.
WILLIAM. Who sets them on?
BAINARD. Fitz Hammon, with De Lacey.
WILLIAM. Bring me the culprits.

Exit BAINARD.

Scene vii

WILLIAM *alone*.

WILLIAM. This spirit must be crush'd:
Treason, when kindled in a camp, will blaze
To flames that spread devouring Death around
If but a spark take fire. This graceless Flambard
Betrays my interest for De Warrenne's: doubtless
De Warrenne in his turn will pay him back.
Such is the friendship kings from Flatterer's find! –
Would I had better brook'd Chester's stern honesty!

Scene viii

WILLIAM. BAINARD. FITZ HAMMON. DE LACEY.

WILLIAM. Fitz Hammon and De Lacey, shameless knights!
Why have ye instigated this commotion?
FITZ HAMMON. My lord, we but lamented noble Chester,

And the whole camp resounded with alarm.
DE LACEY. He was our friend, our Patron, and our Chief –
FITZ HAMMON. Could we behold him doom'd to infamy
And in ungrateful silence sink his praise?
DE LACEY. Has he not merited our ardent zeal? –
FITZ HAMMON. O mighty king! permit us at thy feet –
WILLIAM. Retire! and send me Flambard.
Exeunt FITZ HAMMON *and* DE LACEY.

Scene ix

WILLIAM *alone*.

WILLIAM. I must o'er look
Suspicion and resentment, to consult him.
His Brain is fertile, and his Heart is hard;
So he suggest but riddance of these troubles
His conscience will not scruple ways and means.

Scene x

WILLIAM. FLAMBARD.

WILLIAM. Flambard, thou hearst this cry 'gainst public Justice:
What's to be done?
FLAMBARD. It must be sooth'd, my liege,
By grant of Chester's life, till the Duke's downfall,
Or general Peace, disarm his fierce adherents.
The troops may then endure, what now they brave,
Extreme controul.
WILLIAM. And they shall have't with interest.
But go! repeal their precious Chester's doom
Till times are fitter.
FLAMBARD. Still his rash confederate
The daring De Belesme, may safely fall:
'Twill quicken the surrender of the Garrison
By desperate panic.
WILLIAM. Ay, ay, away with him!
Haste and proclaim his fate: let the whole Citadel,
And chiefly my obdurate uncle Mortaign
Hear that this Hour he dies a Traitor's Death.
FLAMBARD. It shall be done; – but, sire, – De Warrenne –

WILLIAM. Ay, –
De Warrenne, – true, –
FLAMBARD. Since Chester yet must live
May he not hope your Highness' gracious promise
Again will be confirm'd?
WILLIAM. Ay, – let him have her. –
FLAMBARD. Still to your Grace he makes his humble suit
That, as he fears rejection from the maid,
Instant profession of herself and portion
May form th'ostensive terms of Chester's life.
WILLIAM. So be it then; but let me hope, in turn,
His services may grate mine Ears no more.
FLAMBARD. My liege, the pregnant moment for this purpose
T'impress the timid Adela with fear
Is now, when Death, hov'ring o'er De Belesme,
Will scare her quick consent, for Chester's safety.
WILLIAM. Ay, send her hither, with her Rebel Father;
And, ere she learn the traiterous troops protect him
Give her to think 'tis her's to kill, or save him.
FLAMBARD. There is no other way.
WILLIAM. Hither bid, also,
This most impatient BrideGroom: let him witness
My zeal to give him his requested care,
A scorning Beauty for his latter Days.
I joy I share not such felicity.
FLAMBARD. Be but her dowry paid, he'll risk aught else.
WILLIAM. I'll see it done: she shall herself behold
Her Father's danger through her doughty Champion's:
Call therefore forth this hardy De Belesme;
Summon them all! and let thy quick dispatch
Announce alike my vigour and revenge.

Exit FLAMBARD.

Scene xi

WILLIAM *alone*.

WILLIAM. My mind recoils against De Warrenne's claim: –
Yet if these murmuring Soldier's still protect
Chester from Death, attainder, confiscation,
'Tis better with a portion of his lands
Than my own hoards, to pay the call rapacious.

V xi 3 *attainder*: extinction of all civil rights, including the right to property and to transmit or inherit property.

De Warrenne's craft requires no admonition
To shew him in how short a space this dowry,
In common with all Chester's rights, must fall
The penalty of treason, to my coffers.
I understand his speed, howe'er he blandish it
With quaint conceits of Bridegroom eagerness.

Scene xii

WILLIAM. FLAMBARD. DE WARRENNE. BAINARD. CHESTER *with* GUARDS. DE BELESME *in bonds*. ADELA *following* CHESTER.

WILLIAM *seats himself at the upper end of the Tent*.

WILLIAM. Earl Chester, thy desertion merits ignominy
As public as thy crime.
CHESTER. My liege, I pledge
My still unsullied honour for my innocence.
Not as a traitor did I seek the Citadel
To aid in council, or assist in arms:
This hand had sooner stopt the vital stream
That warms this Heart, than rais'd it's sword for Treason.
Disguised, unknown, at peril of my life
I enter'd but to rescue a lov'd Daughter
From ills more fearful to my Soul than Death.
WILLIAM. Most apt pretence! And art thou, too, no Traitor,
Thou, Robert De Belesme?
DE BELESME. Great Heaven forefend!
A Traitor? no! I serve my lawful Prince,
And cast back Traitor on who dare to brand me.
WILLIAM. The fate thou bravst, thou shalt not long solicit.
Bainard, make ready!
ADELA. Oh! –
WILLIAM. Sigh'st thou, lady?
Thou shalt miss no defender. Earl of Chester
One way alone remains to prove thy fealty:
Disown thy new connections with the Rebels,
And instant give thy Daughter to De Warrenne:
So only shalt thou save thy forfeit life.
CHESTER. I cannot play the tyrant with my child:
I do bequeath her to her own decision.
WILLIAM. Then, Adela, hold out thy virgin hand,
Or see thy Father perish.
ADELA. Heaven support me! –

I do resign me to thy kingly power. –

De Warrenne *takes her Hand.*

William. Fair maid, this hand –
De Belesme. O torture!
De Warrenne. How? – what sayst thou? –
De Belesme. Thinkst thou – these Bonds – have nothing galling in them?
That's all, – I mean no more! –
Adela. Alas! – I die! –
O take me hence, De Warrenne! –
De Belesme. Potent Powers! –
Adela. Since I am thine – o hence convey me quick –
Chester. See how she changes colour! – o my child! –
My virtuous child! –
Adela. Haste, haste, De Warrenne, haste! –
I sink! – my strength, my courage –
Chester. O too much!
Can I behold this struggle, – and yet cause it? –
Revoke thy pardon, king! – Renounce, O Adela,
Thy plighted faith! Give me to die at once! –
To die by the quick mercy of the ax,
Not by the woes of this untimely victim!
Adela. No, my lov'd Father, no! – All will be well, –
Live! and a few short pangs will end this strife! –
Live! – that thy Adela may bless her Birth!
De Warrenne. Earl Chester, since the Maiden is content –
Adela. O yes! – and happy! – Take me but away! –
De Belesme. O horrible! –
De Warrenne. What meanst thou? –
De Belesme. Dost thou wonder –
Thus manacled – I thirst for Death? – that's all! –
Adela. Come! – I am ready, – pardon, Lord De Warrenne,
If somewhat faltering, – I shall soon recover, –
Lead me but to the air –
Chester. Stay! stay! release her! –
Horrour is in her Face! – I will not live! –
What has my remnant life to pay this sacrifice? –
Draw back thy trembling hand! – I claim my doom! –
I will not live! – Free be thy soul as pure! –
Rufus, in me alone behold thy prey! –
Come! end the fearful contest! –
Adela. Cruel Father! –
O list not to him! make me not his murderer! –
Come, come, De Warrenne! –

CHESTER. Take her at thy peril! –
She shall not go with thee! – Give, give me Death! –
DE WARRENNE. Heed him not, lady, –
ADELA. O that this hateful dower
From me divided – Royal William, hear me! –
O hear thy suppliant ere thou sealst her doom! –
Permit that in some convent's lone retreat,
Plac'd by thyself; guarded, secur'd, restricted
By thy own mandate – o permit that there
For life confin'd, I wave these nuptial rites,
And leave to thy arbitrament my portion, –
While yet my Father lives – and lives to serve thee! –
WILLIAM. Flighty extravagance!
ADELA. O join, my Father!
If thou wouldst wish me to taste peace again,
If thou wouldst rescue me from endless Tears,
Plead, aid, implore, – render me not a paracide! –
Live, live thyself! – and dedicate thy Adela
To cloister'd solitude, with mental freedom!
CHESTER. Can I resist thee? – No! – my spotless Child,
Cherish thy filial joy! – Thou savst thy Father! –
Bend, bend, o monarch, to our mutual prayer!
WILLIAM. How? favourst thou the wild, unnatural project?
CHESTER. Her tender duty conquers me. O King
To holy vows, not hated rites, decree her,
And let her dowry be the Treasury's spoil.
WILLIAM. The Treasury? – Ha! –
DE WARRENNE. My lord, thou canst not hear –
WILLIAM. Disturb me not. –
CHESTER. The full and ample dowry
I destin'd for her marriage, shall be forfeit.
WILLIAM. Her dowry? – humph! –
DE WARRENNE. My lord! –
WILLIAM. Art thou aware,
Young maid, that there encas'd, to avert all fraud
No Eye uncloister'd must behold thee more?
ADELA. I am most content.
WILLIAM. Not ev'n Earl Chester's?
CHESTER. Oh!
ADELA. Whate'er the terms, I am thankful.
DE WARRENNE. Good my lord –
WILLIAM. Since, then, to prayer thou wouldst devote thy life –
DE WARRENNE. Hold, gracious Sir! – Is not the damsel mine?
And mine her dowry?
WILLIAM. Wouldst thou have me, then
Force her from holy vows?
DE WARRENNE (*apart to* FLAMBARD). Dissembling art!

This 'tis to serve a tyrant!
WILLIAM. I accede!
Since 'tis your joint request, from zeal devout,
The maiden's pious wishes shall be granted.
And now no more! Fulfil your several fates.
Thou, Chester, for a while, must well be guarded:
Thou, Damsel, shalt in safety be convey'd
To learn new lessons in Monastic schools:
Mountjoy, escort her to the neighbouring abbey:
Thou, De Belesme, facing thy vaunted party,
Go, and fill up the measure of thy heroism!
Go, treat thy friends with seeing thy last gasp!
Away! Time presses. Guards, perform your offices.
ADELA. Heaven keep your Highness, then! – yet – ere I go
Where, from this time, all sublunary joys,
All earthly interests, sorrows, or delights,
Must fleet before these Eyes, unstay'd, unheeded,
May I not once, before the awful moment, –
O King! – O Father! – in this final parting
Lighten my loaded Heart with one last vent?
DE BELESME. Who could refuse?
CHESTER. O hear her!
WILLIAM. Well, be brief.
ADELA. First, then, De Warrenne! thou has scap'd an evil
Hardest and heaviest 'gainst domestic peace,
A thankless Wife! one, who with secret horrour
Repugnant still had met thee; – one, whose duties
Had all prov'd tasks; whose daily avocations
Internal agony! while her lonely Hours
With pining discontent had worn her fast,
Blighted her Youth, and blasted thy last Days, –
For, – hence, vain blushes! – thou hadst had a Rival
Whose Image would have canker'd all within! –
Whose virtues would have darken'd all thy kindness;
Whose noble patience in the ills she wrought him
Deep sinking in her soul, had marr'd its peace,
Made happiness an alien to her breast,
And harrass'd her short sleep with mournful visions! –
Such would have been thy Wife!
DE WARRENNE. Amazing hearing!
DE BELESME. O Heaven and Earth! O shackles! Manacles!
Now, now ye bind indeed!
WILLIAM. Enough, enough!
Guards, lead her on at once.
CHESTER. Oh Heaven!
DE BELESME. Yet stay!
Spare De Belesme one short and solemn moment,

Celestial Adela! These Eyes behold thee
Already sainted; and in a few brief seconds
They may survey thee from a purer height:
May I not then presume –
ADELA. O De Belesme,
Be quick, – and probe not deep! –
DE BELESME. Chaste Adela!
O not for Empires would I shake thy firmness.
I mean but to adjure, in yonder cell,
Where holy vows will consecrate thy life,
And soon obliterate each human woe,
One gentle orison of peace, to bless
And sooth my hovering Spirit.
ADELA. O De Belesme!
The prayers of Adela – her humble offerings –
Thy noble deeds – thy woes – thy end untimely –
O De Belesme! require me not to speak! –
DE BELESME. I will not, – virtuous maid! – Heaven bless thy footsteps,
And Hope sublime support thee! –
WILLIAM. No more, no more!
ADELA. Yet – yet – one moment, sire!
WILLIAM. Prithee then speed it.
ADELA. To Thee, my Father! –
CHESTER. Oh! my Heart will break! –
Do not take leave of Me! – My Child! – my Child! –
Turn from me quick! –
ADELA. O yet – one parting blessing! –
CHESTER. One thousand! – O my white-soul'd Adela! –
I cannot speak, Go! yield thyself to Heaven!
ADELA (*kneeling*). O softest Peace, the mede of spotless Virtue!
Look down on this lov'd parent! sooth him! guard him!
Shed on his reverend Head thy sacred balm,
Tell him his Adela retires to rest,
Not to affliction; say she goes content
That she may pass her Days in soft remembrance
Of all his wond'rous goodness, patience, sweetness,
His mild indulgence to her early errour,
His noble precepts for her rules of life.
Tell him his last kind act from sorrow chac'd
It's keenest arrow; grave on his generous Heart
Her thanks, her gratitude, her filial reverence –
O Heaven! (*rising*) he weeps! – my Father! my lov'd Father! –
CHESTER. No more! – I cannot bear it! – turn thee from me –
My victim child! – my guiltless Adela! –

V xii 142 *orison*: prayer.

WILLIAM. Part them all quick! – Bear them their several ways!
DE BELESME. O ruthless William! –
ADELA. Farewell! – a last farewell!

DE BELESME *is carried off by* BAINARD *and* GUARDS *at one side, and* ADELA *by an* OFFICER *and* SOLDIERS *at the other.*

Scene xiii

WILLIAM. CHESTER. FLAMBARD. DE WARRENNE.

CHESTER. Why will the light not close upon these Eyes?
DE WARRENNE. My lord –
WILLIAM. I cannot hear thee now.

Exit DE WARRENNE.

Flambard!
Let me have notice when the blow is struck.

Exit FLAMBARD.

Enter an OFFICER.

OFFICER. A Truce, my lord the king! Earl Arundel
This instant brings a message from Duke Robert
That craves immediate hearing.
WILLIAM. Let him enter.

Exit OFFICER.

He's well arriv'd to wish his son good night.

Scene xiv

WILLIAM. CHESTER. ARUNDEL. OFFICERS.

ARUNDEL. O King! – my son! my son! – issue thy mandate –
WILLIAM. How now, Earl Arundel! art raving mad?
Thinkst thou one Traitor shall another save?
ARUNDEL. O grant him but a moment! – While I speak
Perchance he dies! – send but the shortest respite –
I ask no more! –
WILLIAM. Well, – give him a few seconds, –
If yet he stand my Debtor for disloyalty.

Exit an OFFICER.

Now, Earl, what is thy payment for this clemency?
ARUNDEL. A life of gratitude! –
WILLIAM. That may be short;
But what says Robert?
ARUNDEL. Now, sire, I can speak.
The noble Duke, griev'd at the ills of War,
Wishes to spare th'effusion of more blood
Ev'n at the proper cost of his own interest.
WILLIAM. How knowst thou this?
ARUNDEL. A messenger from Normandy
Has vested the Earl Mortaign with full powers
To offer Peace, and general amnesty.
CHESTER. Listen, O King!
WILLIAM. Well, – what would he propose?
ARUNDEL. Each Baron of each country to enjoy
His lands on each: all captives to be free
On either side.
WILLIAM. And on what terms?
ARUNDEL. The noblest!
Content with his hereditary patrimony
Of Normandy, he yields you England's Throne.
WILLIAM. Then he's my Brother – and we war no more!
ARUNDEL. Now, now, O monarch, might my pardon'd son –
WILLIAM. Ay, – bring in De Belesme.

Exit an OFFICER.

Chester, we're friends!
Recall the lady Adela.

Exit another OFFICER.

CHESTER. Blest hearing!
Auspicious moment! O sire! Faith and loyalty
Be all thy own! and Peace, sweet, heavenly Peace
Smile once again on this War-harrass'd shore!

Scene xv

WILLIAM. CHESTER. ARUNDEL. BAINARD. DE BELESME.

ARUNDEL. My son! my valourous son! my pride! my glory!
Snatch'd from the Grave – O sire! forgive a rapture
Defying all restraint, – Nature's own offspring.
Here, De Belesme, make now thy humble homage
In thy Duke's name, to England's lawful Sovereign.

DE BELESME *kneels*. WILLIAM *raises him*.

WILLIAM. Serve me as thou hast serv'd the Duke my Brother;
I ask no other retribution. Chester,
I see thy wish; thy Daughter is thy own.
Secure me this young Soldier in my Wars,
I'll find some other Heiress for De Warrenne.
DE BELESME. Thou bindst me to Thee now, for life, or Death! – –

Scene xvi

WILLIAM. CHESTER. ARUNDEL. BAINARD. DE BELESME. ADELA. OFFICERS.

CHESTER. She comes! my child! my Adela!
ADELA. My Father!
O mercy most unlook'd for! –
DE BELESME. Blest deliverance!
CHESTER. Can it be given me, then, to fold once more
My bosom's joy in these paternal arms?
ARUNDEL. Propitious Hour!
CHESTER. Come hither, De Belesme!
And here, for once, let Earth behold the junction
Of Happiness with Virtue! Take her; guard her; –
I never saw the one I thought her Peer
Till I knew Thee, thou gallant De Belesme!

Elberta

Sketch for a Tragedy

ELBERTA

Elberta, which Burney never completed, survives in a series of autograph fragments in the Berg Collection. These fragments are numbered from 1 to 302 in a modern hand, probably that of Adelaide Smith, cataloguer of the Collection when the Burney papers were acquired in 1941.[a] The number assigned to each fragment apparently follows the order in which they were found when acquired for the Berg Collection; no attempt seems to have been made to rearrange the jumbled pieces. There is a 60a and an unnumbered fragment (between 47 and 48) but no 270, so that the total number of fragments is 303.[b]

Several of the fragments in the series have nothing to do with *Elberta*, and are omitted from this edition. Among the extraneous material are jottings on *Edwy and Elgiva* and *Hubert De Vere*, and a few draft scraps of what seems to be another historical tragedy. There are also such oddments as El. 1 (containing markings, not in Burney's hand, unrelated to the play), El. 3 (an advertisement for 'Weldon's true Scouring Drops'), El. 92, 150, and 246 (letter fragments), and El. 293 (a chart). The resulting confusion, making an already incomplete and fragmentary work seem far more chaotic than it is, has led Burney's critics to some false assumptions about the play.

In her article 'Fanny Burney: Playwright', Joyce Hemlow declares that *Elberta* 'does not seem to have been carried beyond preliminary jottings' (p. 181), a comment repeated in her biography of Burney (p. 220). Marjorie Lee Morrison, who attempted to unravel the plot, states wrongly that 'it is impossible to disentangle the events of Acts IV and V, and the ending is not supplied' ('Fanny Burney and the Theatre', p. 104). Margaret Anne Doody also attempts to recover the plot, but she too writes that the play exists only as 'fragmentary notes' (*Frances Burney*, p. 180). Among the fragments there are many such notes – such as pieces of plot summary, abstracts of acts and scenes, descriptions of characters, etc. – but there is also a considerable quantity of actual verse, comprising the greater part of a draft of the whole play.

In this edition, Stewart Cooke has created a consecutive text of *Elberta* for the first time from the surviving fragments. The verse has been printed separately and organized by its apparent place in the plot, rather than by

[a] See Joyce Hemlow, *A Catalogue of the Burney Family Correspondence 1749–1878* (New York: New York Public Library, 1971), p. xiv.

[b] Number 270 may have been a letter of 2 April 1796 to Burney from her father. A note on the inside cover of the box housing the scraps states that this letter has been removed.

the cataloguer's numbers. For the fragments constituting Acts one to three, Burney provided both act and scene numbers. The Act four fragments have scene numbers up to scene ten, while only a few of the Act five fragments have even act numbers; for Act four, scene eleven to the end, therefore, scene numbers are editorial insertions. Most of the remaining material has been printed in the form of textual notes. In addition an appendix is provided, containing scenes from *Elberta* that are either too general to place or provide possible plot developments conflicting with those in the text. While the newly reshaped play still has many obvious lacunae, it is much closer to completion than had previously been believed. *Elberta* can now be read as a whole and fruitfully compared to Burney's three previous tragedies.

One of the *Elberta* fragments has featured prominently in the few critical studies of the play:

> A Female is mentioned, who wild and unknown is seen roaming about – no one is informed whence she comes – woe is in her voice, terror in her aspect, – she never weeps, yet frequently wails, tho' in terms unintelligible from their wildness –
> Her interesting appearance –
> Some she affrights – others is derided by –
> Her fierce harangues, though wild when offended –
> Her gentle supplications to shadows.
> Her inattention to pursuit –
> Her stated calls on.
> Stated visits to. (El. 265; *Elberta*, Appendix 10 below)

Part of this fragment was first quoted by Joyce Hemlow in her article on Burney as playwright (and later in her biography) in order to characterize the play as a whole, in which 'no scrap promises any brighter end than gnawing famine, violent deaths, and madness' ('Fanny Burney: Playwright', p. 181; *The History of Fanny Burney*, p. 220). Marjorie Morrison, similarly, regards this fragment as one of two that provide the 'genesis of the play'. And Margaret Anne Doody uses the same fragment to argue that 'the plot was evidently designed to show Elberta's progressive breakdown into madness under the pain of privation. . . . the heroine evidently becomes a mad wanderer, like some of Wordsworth's early characters' (*Frances Burney*, p. 192). In the text of the play itself, however, as opposed to this single fragment, Elberta does not go mad and is not unintelligible to other characters. She does not wail or supplicate shadows; these seem to be plot developments that Burney considered but rejected. Thus, since this fragment is one that conflicts with the reconstructed text, it is placed in the appendix to the play below (p. 306).

Burney began writing *Elberta* on 5 June 1791, only a month before the end of her five years of court service. In a diary entry she relates her composition of the play to her mental distress at the time: 'So melancholy, indeed, was the State of my Mind, from the weakness of my frame, & the never-ending struggles for the rest I sighed after, that I was never alone but to form scenes of 'foreign woe,' when my own disturbance did not occupy me wholly. I began – almost whether I would or not – another Tragedy!' (*Diary and Letters*, IV, 478; Berg ms. 4327). In an unpublished passage, Burney contrasts *The Siege of Pevensey*, which 'was not dismal enough', with *Elberta*, a work of 'deepest Tragedy, which first had occurred to me in the worst part of my illness in January, but which I had not thought of since my quitting my room' (Berg ms. 4327). There are no further references to *Elberta* in Burney's letters or journals. In October 1791 she refers to work on her third tragedy, *The Siege of Pevensey*, and notes that *Edwy and Elgiva* and *Hubert De Vere* are complete in draft form, but she does not mention *Elberta*.

Doody states that Burney returned 'briefly to *Elberta* after her father's death' in 1814 – and after the publication of her final novel, *The Wanderer*, in the same year (*Frances Burney*, p. 384). Some of the manuscript pages are taken from the September and October sections of an 1814 appointment book, suggesting that Burney may have written part of the draft of *Elberta* in that year. Other manuscript pages are the backs of letters written between December 1785 and January 1811.

PERSONS OF THE DRAMA

OFFA Commander of Mercia
ARNULPH Nephew and Heir of Offa, and commander of the King's Forces
SAXRED Minister confidential of Offa
WILFRID Confidential servant of Arnulph
CEOLRIC General of Mercia, under Arnulph
INFORMO
ENVOY
ELBERTA Captive Daughter of the late Ethelbert, King of East Anglia, slain treacherously by Offa, and sister of his deposed son, Ethelric, and wife of Arnulph
ACCA Nurse to her Children
ROWENA Cottager
COTTAGER
SOLDIER

Mercia: the largest kingdom of the heptarchy or seven Saxon kingdoms. According to Hume's *History of England*, Mercia consisted of all the middle counties of England and was so named because it bordered on all the other six kingdoms and Wales (I, 40). By the middle of the eleventh century, the time of the play, the heptarchy had given place to one united English kingdom, and Mercia was one of the four principal earldoms.
East Anglia: the kingdom whose territory roughly comprised what are now the counties of Norfolk, Suffolk, and Cambridgeshire. East Anglia was also one of the four principal earldoms by 1066. Hume relates that Offa, King of Mercia, treacherously murdered in 792 Ethelbert, King of East Anglia, who had come to seek the hand of Offa's daughter, Elfrida (*History*, I, 41). It was, perhaps, the description of this incident that prompted Burney to name Elberta's father Ethelbert and his murderer Offa.

Persons of the Drama: pieced together from El. 55 and 61.
ARNULPH: the ms. reads 'Ardulph' here and occasionally after.
CEOLRIC: El. 61, a list of characters, describes Ceolric as 'hot – blunt – hasty. generous. compassionate. liberal Hea‹rted›'.

ACT I

ACT I: several fragments provide the exposition. The fullest of these, El. 292, follows:

Ethelbert, protector and favourer of Edgar Atheling, was defeated at the Battle of Hastings,[1] from which time he lived in Exile, his Title of[2] forfeit, his person exiled, his estate confiscate, and his family banished.

Arnulph, son of the potent Baron,[3] in the wars meets them, falls enamoured of Elberta, his Daughter, privately marries her, and keeps her in retreat five years, with two Children, visiting her in secret. Her person unknown, and all around too much loving her to betray suspicions.

Arnulph, called now from Northern command, leaves her; accidentally wandering alone, she is encountered by Informo, a wretch who had deserted from her Father, on his failure, and who had served Offa, the successor to Ethelbert's honours and estates. He knows, and seizes her – and carries her to Offa. Offa, becoming enamoured, would marry her. She refuses him; but he owns not possessing her, as a general amnesty is expected, and he fears refunding her property, should she marry and have Children. Otherwise, it is all entailed on the male Heir, except her Mother's dowry.

In El. 23 and El. 263, Elberta is the daughter of Elwolf, chief of staff of Edgar Atheling. In El. 23, El. 24, and El. 25, she has only one child. In El. 24, her servant's name is Githa. Arnulph, in El. 23, is 'Chief of William's troops over the Insurgents for Edgar Atheling'. In El. 27, Elberta is captured, not by Informo, but by 'a party, who carry her off to Offa – William's Chief'.

An unnumbered fragment, deposited between El. 47 and El. 48, explains why Elberta's children were not also taken:

The deserter, not knowing her children, nor suspecting her marriage, left them with Acca, to whom she had sent to [an illegible word] them by Wilfrid's bounty.

El. 28 provides a summary of Offa's actions after Elberta's refusal of his marriage proposal:

Offa falls in love with his Captive – and resolves to demand her on the amnesty and the Casket as her dowry.

Amnesty announced.
He offers.
Is rejected.
[an illegible word] and irritated.
His confidant brings other papers – of [an illegible word] address –
Deems them treason and resolves to detain her.
Offers marriage to save.
Rejected again.
Threats Treason revealed.
Trembling – for Arnulph.
Yet firm in innocence – and [an illegible word] disdainful –

El. 26 summarizes Elberta's reaction to Offa's threats:

In the accusation.
Paper produced.
She starts –
Tis a letter of Arnulph.
Durst not avow.
Seems conscience struck.
Yet asserts innocence.

[1] 'and thence saved by a flight to Malcolm of Scotland' is deleted.
[2] Burney left a blank space, which was never filled in, for the title.
[3] an insertion, replacing 'a young Baron'.

Scene i

OFFA, SAXRED.

A messenger from Court demands ELBERTA. *Confusion*!

OFFA. Which way have they trac'd her? Speak!
What secret miscreant, What officious traitor,
What base insidious spy proclaim'd her here?
Who brings this villainous news?
SAXRED. Envoy.
OFFA. Her Father's friend – my very deadliest foe!
Why he will sneer with joy
Exult
Much I suspect the king is mov'd to favour her –
Still, as Heir male, then

Enter ENVOY *to* OFFA *and* SAXRED.

ENVOY. Joy to the fair Elberta of East Anglia!
Joy to the gentle Roof that gave her refuge.
Peace, smiling Peace disbands our Hosts with honour
And conquering William spreads his mighty arms

I i 3 *insidious spy*: Wilfrid. See below, Act 1, Scene x, summary (El. 45).
I i 13 *conquering William*: William, Duke of Normandy, who defeated Harold Godwinson, Earl of Wessex and King of England, at the Battle of Hastings, 14 October 1066. He was crowned King William I on Christmas Day 1066.

Scene i: El. 58, which provides a synopsis of each scene of Act I, describes Scene i as follows: 'Envoy arrived from Court to Elberta'. El. 4 summarizes it as: 'The Messenger's announce of Amnesty and Freedom for the Daughter of the Chief of Harold's staff, Elberta'. El. 5 provides a lengthier summary:

> Saxred announces an Envoy from court to Elberta. Alarmed – fears general amnesty from Atheling's subjection or submission. Resolves she shall not appear – charges Saxred to assist.[1] The Envoy his particular Enemy, and the friend of her dead father. Saxred says at least her father cannot arise from the Grave to re-seize the property. While debating[2] Informo enters to say Envoy cannot wait – forced to admit him. Agitated and disordered Spirits.
>
> [1] an illegible word is deleted.
>
> [2] 'hurried by envoy forced to admit [an illegible word] – disorder' is deleted.

I i 1 s.d. *demands* ELBERTA: the ms. reads 'demand Elberta?'.
I i 1 *Speak*: an insertion, replacing an illegible word and 'Why'.
I i 4 *Envoy*: the ms. reads 'Envoya'. Elsewhere the ms. sometimes reads 'Envoya' or 'Envoi'.
I i 1 s.d.–9 OFFA, SAXRED ... *Heir male, then*: from El. 50. The speakers' names are editorial insertions.
I i 9 s.d.–17 *Enter* ENVOY ... *or clemency*: from El. 251.

By Glory led, to shelter, not assault
The Partisans of Edgar Atheling,
Who, yielding to the Fortune of the Day
Resist no more his power, or clemency.
OFFA. I hear with due respect my sovereign's mandates;
But . . .
ENVOY. And feelst thou nothing warmer than respect?
No animating glow? no genial pleasure
That Mercy, raising high her fostering head,
Receives the helpless, harbours the unhappy,
Forgives the penitent, arights the injur'd,
And from dark Persecution's vengeful rage
Rescues the Land to mild and equal Justice?
OFFA. See you not I rejoice?
ENVOY. I crave your pardon.
I thought you sad. Yet tydings such as these,
That renovate bright hope in sorrowing beauty,
And raise a trembling virgin from despair
Might win to sympathy the gloomy anchorite,
Still Envy, and surprise even discontent
T'unbend her hardening muscles with a smile.
But werefore see I not the gentle lady?
OFFA. Saxred, I pray you summon her. Allow me
Some little question. Is this amnesty
Held forth to All? or comes its partial grace
But to the favour'd Daughter of Earl Ethelbert?

I i 15 *Edgar Atheling*: the aetheling, a prince of the royal house of Wessex, had, as the grandson of Edmund Ironside, a stronger kinright to succeed to the throne than did William. The leading citizens of London, led by Ealdred, archbishop of York, and earls Edwin and Morcar, 'elected' him as king following the Battle of Hastings. By mid-November, however, Edgar and his supporters submitted to William at Berkhampstead. Hume relates that Edgar escaped into Scotland in 1068 where he was received by King Malcolm III, who gave refuge to any enemies of William. In 1071, however, 'despairing of success, and weary of a fugitive life', he 'submitted to his enemy; and receiving a decent pension for his subsistence, was permitted to live in England unmolested' (*History*, I, 210). It is after this submission that the action of the play takes place.
I i 31 *anchorite*: or anchoret; a recluse, hermit.

I i 22 *raising high her*: 'spreading wide her' is given as an alternative.
I i 22 *head*: 'hand' is given as an alternative.
I i 18–28 *OFFA. I hear . . . such as these*: from El. 252.
I i 29–33 *That renovate . . . with a smile*: from El. 253.
I i 34–38 *But werefore see I . . . Earl Ethelbert*: from El. 291.

Scene ii

OFFA, ENVOY.

Banishment and confiscation repealed
By potent intercession

ENVOY. O pardon me!
I had forgot to you the joy is naught –
But yet I knew her Father –
The general amnesty
Still, as Heir male –
Though sure – the lovely Elberta may marry –
And – when no longer Childless . . .
OFFA. Nay – the Inheritance is still my own
The will awards it to Heirs male alone
Such the entail
Yet, but of treasure, money, Jewels, goods
Some little matter may, perchance, be found
Her mother's dowry is her settled right.

I ii 10 *entail*: the settlement of the succession of a landed estate so that it cannot be bequeathed at pleasure by any one possessor (*OED*). Because Ethelbert's estate has an entail male, Elberta cannot inherit it.

Scene ii: El. 58 summarizes Scene ii as 'Envoy brings amnesty to Elberta'. El. 14 provides a longer summary:

> He comes to bring News to Elberta of amnesty to all parties by Edgar Atheling's submission to the Conqueror[1] and that her banishment and confiscation are repealed by potent intercession. Offa says Estate vested in Heir male, therefore still his, her Father is dead, and she single. Envoy says treasure and [an illegible word] are hers. And all, if she marries and has issue.[2] Saxred goes for her, announces she is ill. Envoy says he will return next Day.
> She was Daughter of dearest friend –
> Entitled to her mother's dowry –[3]

El. 6 verso reads: 'Saxred enters with message Elberta is too much indisposed to be seen. Envoy suspiciously wonders – says he will call again next morning'.

[1] 'Edgar Atheling's submission to the Conqueror' is an insertion, replacing an illegible word followed by 'and Stephen'. Stephen was king of England from 1135–54. Burney must have intended at an early stage to set the play during the anarchy that characterized Stephen's reign.

[2] 'They spar' is deleted.

[3] 'Offa sends off Saxred for Elberta' is deleted.

I ii 1 s.d. *intercession*: two lines are deleted:
Joy to the fair Elberta of East Anglia
Joy to the gentle House that gave her refuge.

I ii 2 *naught*: El. 294 contains the following dialogue:
OFFA. Sure he will ne'er forget that I the first
Avow'd his right, and cancell'd Athelings?
ENVOY. I had forgot – To you the joy is naught
Save in the liberal spirit that partakes
The good of others –

I ii 1 s.d.–13 *OFFA, ENVOY . . . settled right*: from El. 16. The speakers' names are editorial insertions.

Alas, Sir, attainders are eternal Causes
Not merely in their sufferance, but repeal –
ENVOY. Tis true
Her noble Father cannot be recalled
Death, in dishonoured silence, Exile, sadness
Cut short his length of Days –
OFFA. In these times of civil war
An honest man is still the prey of accident.
He knows not where to seek for peace or glory –
Him whom he serves to Day with every honour
To morrow is as nothing –
ENVOY. Yes – but he who pursues
The path of Honour, is sure of Fame:
For Fame nor hangs on Interest nor on Life.
'Tis true
Honour – Celebrity hang on Success –
But Fame, fair Deathless Fame, is Honour's mede,
Prosperity it mocks, and braves Adversity,
Still to the Right undeviating it hies,
Scoffing disaster.

Scene iii

OFFA, SAXRED.

SAXRED. There's little chance if now, while poor and friendless she dared resist –

I ii 14 *attainders*: extinction of all civil rights, including the right to property and to transmit or inherit property. Offa is arguing that the amnesty does nothing to change the fact that Ethelbert could not transmit by descent, because he had been outlawed for treason.
I ii 29 *mede*: or meed; wages, recompense, reward.
I ii 31 *hies*: hastens, speeds.

I ii 14–20 *Alas, Sir ... length of Days*: from El. 15. The speakers' names are editorial insertions.
I ii 21 *accident*: an insertion, replacing 'William'.
I ii 26 *Life*: 'Fair Fame is Honour's mede – ' is deleted.
I ii 29 *mede*: 'But Fame is Honour's mede – ' is given as an alternative line.
I ii 20–32 *OFFA. In these times ... disaster*: from El. 250. The speakers' names are editorial insertions.
Scene iii: El. 58 summarizes Scene iii as 'Resolve to press Elberta for marriage'. El. 56 provides more details:

Alarm. Rage. Determine not to tell Elberta – but try once more to obtain her hand. If he fails, he must[1] convey a casket of peculiar value to a Tower he commands.[2] Saxred undertakes it. Fear she will now seek other alliance. Owns when Informo first told him she was in England he meant confinement and impeachment She and all being banished – but seeing her beauty, and dreading amnesty, resolved marriage, not doubting acceptance in such a plight.[3] Say if she persist, it must be discovered she is not ill – to prevent her marrying first essential. Resolve any way to obtain her, and prevent that, if possible.[4] He

She is sullen – silent – sad – mysterious –
She now may elsewhere hope a settlement as honourable.

OFFA. I thank your honesty –
I hate that fellow! – My dear Saxred
Perchance for her release she may give way.
If she still resist
What I have now of treasure in the House
I will carry to fortress.
This proud Envoy
May care to claim –
Will't carry it?

SAXRED. Most willingly.

OFFA. Nay – 'tis not much
A trifle – scarce worth sending –

Scene iv

OFFA *solo*.

OFFA. What? money – treasure – jewels – all –
Why have I hoarded –? Why, with niggard hand
Denied even to myself life's luxuries?
I might as well have been proper –
Pass'd in mankind for generous and bounteous
Had I surmiz'd my savings and oeconomy
Thus to have ended! –
To morrow he'll be here!
He's wondrous prompt – most marvellous officious
He means, perchance, to marry her himself.

---

sends Saxred therefore to try to sooth.[5] And announce in form he begs a parley with her. Bids report her look and answer, and try to plead her advantage in acceptance and her danger in denial stronger than in decency he can do either himself.
The dialogue is from El. 57.

[1] an insertion, replacing 'will'.

[2] 'of peculiar value ... commands' is an insertion, replacing 'to a Fortress'.

[3] 'Informo, her Father's old disbanded soldier and [an illegible word] had met, knew and seized her, and brought her prisoner for [one or two illegible words] Offa' is deleted.

[4] 'Saxred advises she may walk in the air, lest enquiry arraign him for confinement after amnesty' is deleted.

[5] an insertion, replacing 'He sends Saxred to tell [two illegible words], strictly ordering no domestic follow or approach. Thinks this will sooth'.

Scene iv: El. 58 summarizes Scene iv as 'Resolves not to trust Saxred but Informo with Casket'. El. 60 provides a fuller summary:

> Ticklish times. Does not like the readiness of Saxred to take casket. He is his faithfullest friend, but a man, therefore insensible to all when self's in question. Will sooner trust Informo, whom he can pay at once. Calls him – Casket contains money and treasure, which the pardon will render Elberta's, if found – if concealed, it may seem purloined before his possession of the Estate –

I iv 1 s.d.–10 *OFFA solo ... marry her himself*: from El. 59.

Tis hard to trust in man –
He's base by nature –

Scene v

OFFA, INFORMO.

ARNULPH *is announced from army.*

OFFA. Bid him wait – what would he –
He is
I gather from the soldier who attends him
Tis public that an amnesty is pass'd
Between the Conqueror and Edgar Atheling
And all adherents
He brings my Summons then to join the army.

I iv 11–12 *Tis hard ... by nature*: from El. 60a.
Scene v: El. 58 summarizes Scene v as 'Says Arnulph just arrived from Army'. El. 62 provides a lengthier summary:

Informo announces Young Arnulph just arriving[1] from the army: the soldier, who precedes[2] shouting to the Servants d'avance a general Peace.[3] Alarmed lest summoned to Court and in dread how to have her, or expose treasure. Asks Informo to convey a casket to the Fortress. Informo complies for pay. Desires[4] to see Saxred before Arnulph arrives – as it will be suspicious not to name a prisoner of her consequence now her sejour is known – and he must first learn his success, to know what to style her.

El. 63 contains a similar summary:

Offa informed Arnulph is arriving from the Army, with Peace and Amnesty to all awarded by the Conqueror – and to send Offa to Court, and take his command.

Summons Informo to consult about the Casket he had taken from Captive Elberta, the Daughter of the last chief of Edgar Atheling found in a Cave *or* Offa having pillaged Casket in belief it could never be known, as Elberta he thought captive for life – now, distrait – resolves to marry her – and claim the casket – or openly take it – as her dowry –

Sends Saxred to prepare her for an abrupt courtship.

The dialogue is from El. 17. The speakers' names are editorial insertions.

[1] an insertion, replacing 'alighted'.

[2] an insertion, replacing an illegible word.

[3] an insertion, replacing 'amnesty'.

[4] 'Saxred to return before he sees Arnulph, high spirited young officer whom he hates for romance and chivalric character. May wait to settle with S' is deleted.

I v 1 s.d. *INFORMO*. 'Elberta, then, while Childless' is deleted.
I v 5 *the Conqueror ... Atheling*: an insertion, replacing 'K. St. and the Empress Maud'. Maud or Matilda was the daughter of King Henry 1 and wife, first of Henry V, Holy Roman Emperor, and second of Geoffrey le Bel, Count of Anjou, Duke of Normandy. She was a pretender to the throne herself. Her determined opposition to Stephen culminated in Stephen's recognition, in a royal charter issued at Winchester in 1153, of Matilda's and Geoffrey's son Henry as his successor and heir to the kingdom by hereditary right.

Scene vi

OFFA *solo*.

OFFA. I dare not trust that fellow with the Casket
He betray'd His Master.
I trust not hot Romance –
Sooner I'd trust
Fire with my Parchments, Water with
Than Prudence to the hot romance of Youth
If Arnulph comes
Hero! – Captive
I must not leave her to him
No – to old Ceolric I will entrust her –
Yet Ceolric I like not!

Scene vii

OFFA, SAXRED.

OFFA. How shall I leave Elberta?

Scene vi: El. 58 summarizes Scene vi as 'Says he will not trust Informo. Perplexity'. El. 36 adds more detail:

Repents naming casket to Informo a traitor already – perhaps he may seize or inform. – He had already betrayed his 1st. master Ethelbert.

Dislike of Arnulph, a high spirited young Hero, of chivalric and romance propagating Character.

Offa particularly desires to avoid Elberta's meeting Arnulph – a captive Beauty is just the thing to instigate interest and chivalry in a young Hero – and a young Hero is just the thing to catch – soften[1] – the heart of a Captive Beauty – Such liaison though trite since the beginning of the World, will be current to its end.

[1] a word with the initial letters 'cha' or 'chu' was begun here but not finished.

I vi 1 s.d.–2 *OFFA solo ... Master*: from El. 18

I vi 3–6 *I trust ... of Youth*: from El. 34.

I vi 7–11 *If Arnulph ... like not*: from El. 38.

Scene vii: El. 58 summarizes Scene vii as 'Plan perjuries to detain Elberta'. El. 19 provides the following summary:

Saxred says no chance. Offa plans perjuries darkly first hinted, then fully, to detain her, lest she marry elsewhere. Will go to probe and press finally; if not to confine her. Saxred advises her walking out, that papers may be found in her room of treason: While the air allowed prevents suspicion on their part, being her due[1] after pardon; she may be watched carefully, though not molested.

She is mysterious, but firm, says Saxred.

Who could have obtained this officious pardon?

Envoy would be the first – but how learn she was here?

He must immediately to the Conqueror to assure her. difficulty how to have her, since some spy or treachery is certainly here acting for her –

The dialogue is from El. 37. The speakers' names are editorial insertions.

[1] an insertion, replacing 'right'.

If still she's obstinate – Is there no device
That may restrain her?
Nay I but here have found some certain papers
Of her late father, breathing dangerous.
SAXRED. I'll form you some at hand.

Scene viii

OFFA, SAXRED, INFORMO.

OFFA. Come to my Room to settle all – I then
Will see young Arnulph –
But first, tell Elberta
I shall beg a parley with her –
And let her walk uncontrolled – Giving strict order
That no domestic follow or approach –
Then quickly follow Saxred and myself
Let us arrange some plan – that seeing Ceolric
I may be well prepar'd, if he, besides,
Has heard this tattling tale.
You won't object to witness for me
What I may find expedient to produce?

Character of CEOLRIC.

SAXRED. Sir, Ceolric.
OFFA. O pestilence on pestilence!

I viii 4 *parley*: conversation, conference.

I vii 5 *dangerous*: two illegible words follow.
Scene viii: El. 58 summarizes Scene viii as 'Ceolric announced. Elberta allowed to walk. Retire to plan'. El. 42 also provides a summary:

> Informo comes to announce Ceolric. Offa detests him – will arrange Informo that he may announce her guilt to him, and so give it air of truth before he goes off to impeach her at Court. He agrees to witness against Elberta. They go to consult above, and defer seeing the two Knights till afterwards, that the tale may be plausible, and hang together – and that she may have had the air of being first accused. They are to find the papers in her room while she walks.
>
> Plot
>
> The Law of Amnesty freeing Elberta, Offa's fears of her revealing the secret casket determine him to detain her – and he puts papers and conspiracies said to be contained in casket – He gives her in charge – but not the fraud to old Ceolric.

I viii 3 *But first*: an insertion, replacing 'Meanwhile'.
I viii 9 *well*: an insertion, replacing 'all'.
I viii 1 s.d.–12 *OFFA, SAXRED ... produce*: from El. 20. The speakers' names are editorial insertions.
I viii 12 s.d. *Character of CEOLRIC*: 'OFFA, SAXRED, INFORMO' is deleted.

I hate that Ceolric – austere and melting,
Gentle and Stern – he's made of contrariety –
I know not how trust – or how suspect him –
A man he seems at times fit to encounter
Man as he is,
At times, a Boy, a driv'ler,
As if the dupe of fools, though scoff of knaves –
Stiff when offended, as becomes a man
But mawkish when appeased like a weak Girl –
Tell him I'll see him in a second –
Let me arrange some plan first of detention, if
I am summoned – as I can then make her over as a
Prisoner of State herself to his Keep in the Tower he Governs.
I may trust him, though my Enemy.

Scene ix

ELBERTA *sola*.

Changes to a Garden.

ELBERTA. Ah why, since born to woe – should every moment
Of passing time, have power to renovate
Those hopes and views that nourish vain expectance
Those dangerous – dear delusions of the soul
That urge us to forget the appointed tenure
Of short and sad mortality? 'Tis strange
The doom of death, so woven in our frame,
Known with our earliest knowledge, kept in memory
By Friends and Foes alike, that, dropping round us
Spread wide the unchanging Law of Dust to Dust,
Should still seem far from individual man
Though of his race the appendage of Existence!

I viii 15 *contrariety*: two illegible words and 'himself' are deleted.
I viii 17 *at times* – : a second 'at times' is deleted.
I viii 12 s.d.–26 *Character of* ... *Enemy*: from El. 64. The speakers' names are editorial insertions.
Scene ix: El. 58 summarizes Scene ix as 'Hopeless of all: even Wilfrid failing'. El. 22 also provides a summary:
Hopeless of Wilfrid, her last resource, therefore of all. Laments Husband and Children. Sees Ceolric approach. Amazed at liberty. Resolves to address him. Knows him the friend of Arnulph.
I ix 3 *views*: an insertion, replacing 'fears'.
I ix 3 *vain*: an insertion, replacing 'fond'.
I ix 1 s.d.–6 *ELBERTA sola* ... *mortality*: from El. 256.
I ix 6 *'Tis strange*: ''Tis o'er' is given as an alternative.
I ix 6–12 *'Tis strange* ... *of Existence*: from El. 255.

With Me Even Disappointment now subsides.
Wilfrid, my last support, abandons me:
Ah, whither then may poor Elberta turn?
O Husband of my Heart! when may we meet?
When, my deserted Children! may I own ye?

Scene x

ELBERTA, CEOLRIC.

ELBERTA. O Ceolric, one moment lend thy Ear,
  Hear a lost Captive's miseries – and redress them! –
CEOLRIC. What can I do for thee?
ELBERTA. What wouldst thou do

I ix 14 *Wilfrid*: two illegible words are deleted.
I ix 15 *then may*: an insertion, replacing two illegible words.
I ix 13–17 *With me ... own ye*: from El. 81.
Scene x: El. 58 summarizes Scene x as 'She begs his help and tells her Tale. He promises protection'. El. 4 says simply 'The appeal of Elberta for aid from Ceolric'. El. 47 offers two possibilities: 'Elberta tells her tale and reveals the opening Drama either to Ceolric when cast on him by Offa or to Arnulph when parleying with him clandestinely'. El. 45 provides a long summary:

She rushes to beg his help – he asks what he can do? She pronounces her woes: An orphan.[1] Daughter of a banished man, an unacknowledged wife of a loyal Husband, a doating mother of two deserted children – He replies he had obtained her release already. Amazed, she questions. The faithful Wilfrid had told her situation, he had applied to conqu‹eror› and received promise, and her Father's friend Envoy had undertaken the business.

She relates persecution of Offa – and her abhorrence –

She speaks of flight for Edgar Atheling[2] with all her name and confiscation, and Offa succession[3] as nearest kin – Says he came to see if she was released – and will go and demand satisfaction.

She begs him not to be precipitate.[4] She demands his friendship to Arnulph and he replies he will take her to his own care till they may be authorised to live together –

She says marriage must not be avowed till Arnulph obtains the King's Ear for pardon at it's being formed with a banished family without home.

She tells character, office, enterprise of Wilfrid. Forced to serve Offa in House for business, sees her, and finds means to let her know he will assist. Undertakes to bring children to Cottage on the moor – that in their walks[5] she may see them from her high Chamber on the Heath! He carries word to Acca, and that she must call him her own. She expects him every moment. Ceolric despises Atheling, or would have served him! –

[1] an insertion, replacing 'He replies'.
[2] 'flight for Edgar Atheling' is an insertion, replacing 'banishment for Maud'.
[3] 'or enough of these Sc. 2' is deleted. 'or nearest scene' is added in pencil. 'He says he has himself orders from Stephen to free all the state Prisoners from the Tower of which he has the command' is deleted.
[4] 'and hear her. She tells tale of marriage and Children if not told' is deleted.
[5] 'to Cottage ... walks' is an insertion, replacing 'Where'.

I x 2 *miseries*: 'injuries' is given as an alternative.

Hadst thou a Daughter
A Prisoner in her own paternal Castle,
Reft of her Father in sad banishment
Sever'd from every tie of human fondness,
All that affection may unblushing own,
All the chaste duties of the social life
Alien and orphan in her native land
And for what it might bless
Is transfixed to the curse of persecution
The odious passion of disgusting Love
What wouldst thou for a daughter thus oppressed
Hadst thou a daughter
Sever'd from ... all! all! all! –
O let me own the whole!
A tender Husband and two lovely Children!
Busy Honour, and Respect, may follow pride ...
Love and affection wait for softer calls.
Unhappy Arnulph! crown'd with victor wreaths,
He hastes to cast them at the Royal feet
Of him who robs him of their dearest mede!

CEOLRIC. What do I hear?

ELBERTA. Alas! The Tyrant Offa
Tears the dread secret from my soul – ...
His Love untimely ...
Ah think
The Heir of Offa – Mercia's future King –
Thy friend – O dearer tie than all! will thank thee!
How can he hope to win Elberta's Love?
How can he view her cold, averted Eye
And even wish to claim her? –

I x 29 *he*: i.e., Offa.

I x 4 *Daughter*: El. 44 reads:
Hadst thou a friend
What wouldst thou for a friend
Chaced by injustice, terrified by wrongs
Injured by power, ...
What wouldst thou do ... but

I x 5 *Castle*: El. 53 explains, 'The Castle Offa commands had been the inherited mannor of Eric, Father of Elberta'.

I x 6 *of*: 'Servants' is deleted.

I x 8 *that*: 'the chaste' is deleted.

I x 8 *unblushing own*: 'All the chaste duties, all the soft affection' is deleted.

I x 1 s.d.–17 *ELBERTA ... lovely Children*: from El. 52. The speakers' names are editorial insertions.

I x 18–28 *Busy Honour ... thank thee*: from El. 10. The speakers' names are editorial insertions.

I x 29–39 *How can he hope ... by coldness*: from El. 33 verso.

What, with aversion, is the nuptial promise?
CEOLRIC. Thy virtuous bosom, that by Truth abides
In Honour's native plainness, cannot estimate
The wayward bliss of worldly, crafty minds.
The common race of men view youth and beauty
As sole ingredients for the nuptial knot;
Their own choice satisfied, they ask no more:
Unaw'd by distance, unallarm'd by coldness,
They give to young timidity and shame
Trembling repugnance: or even where they discern
Aversion for its root, they still conceive
Time, and adroitness will remove each difficulty:
Or deem that Beauty is alone sufficient.
ELBERTA. Short-sighted such presumption! ill they scan
The delicate female Heart, and ill conceive
It's secret agony, its latent horrour,
It's bitter pangs, from exquisite disgust –

I x 48 *from*: an insertion, replacing 'it's'.
I x 40–48 *They give ... disgust*: from El. 33. El. 21 and 11 give a variant scene; El. 21 contains all but the last three speeches (to 'himself retreating'):

Ceolric says he has heard from Wilfrid of her captivity, and hastened to report it – and hoped to have met her free, as Envoy undertook it.

ELBERTA. Did Wilfrid tell you more?
CEOLRIC. No: he referr'd me . . .
ELBERTA. Hear then . . .
I am Arnulph's wife –

Says he will demand reason she is not free: she stops him to ask for Arnulph –

CEOLRIC. Arnulph you love –
ELBERTA. I love him now no more –
CEOLRIC. Heavens! werefore this dire change!
ELBERTA. He's not the man I thought him – I detest not
He who emprizes what is termed imprudent,
'Tis mostly generosity and nobleness –
But he who follows wild his wish of youth
His fervour of romance, and risk of ruin –
And, once engaged, fails and becomes faint hearted
Involving others – and himself retreating –
CEOLRIC. By Heaven if he'd wed – and leaves thee thus
He merits . . .
ELBERTA. Alas compulsion – to defend his Country –
CEOLRIC. Was't that? – He ever was a noble Boy
From his first Youth –
I gave it into charge of cautious Envoy –
He loved thy Father – and

Scene xi

ELBERTA, ARNULPH.

ELBERTA. O Arnulph! my loved Arnulph! is it you?
ARNULPH. Rejoice you so to meet me?
ELBERTA. Rejoice I? Where must be . . .
When I rejoice not?
ARNULPH. Cruel Elberta! –
ELBERTA. If that be cruel –
Elberta never can be kind!
ARNULPH. Well then rejoice – my wife! my gentle wife!
That my sunk soul may animate once more
Nor can I see thee happy – yet be wretched?
ELBERTA. O Joy of Joys, O blissful hallow'd moment –
Have I known sorrow? – never –
Was the loved lord of every hope and fear
Torn from my arms –
Have I been reft of my loved little ones?

Scene xi: El. 58 summarizes Scene xi as 'Ceolric unites and promises them assistance'. El. 2 gives a slightly longer summary:

Ceolric, seeing Arnulph, calls to him – he starts with amaze joy – fear – at sight of Elberta. She sinks with joy and confusion and timidity: he is retreating fearful; but Ceolric bids them embrace and leaves them, promising protection – .

At the beginning of the fragment, 'Ceolric, seeing Arnulph' and 'Arnulph approaches – or Ceolric fetches him' are deleted. At the end, an illegible line is deleted.

I xi 1 s.d.–8 *ELBERTA ... be wretched*: from El. 65. The speakers' names are editorial insertions. The fragment is marked 5.2, but it clearly does not belong there.

I xi 9–13 *O Joy ... little ones*: from El. 43. El. 51 provides a variant scene:

Arnulph comes, Elberta flings on her veil and retires –

ARNULPH. Who's that?
CEOLRIC. A Captive – n'importe
ARNULPH. Poor lady! – Suffer, perchance a soldier's wishes – pity – That air! –

She waves her hand in silence and declines or meeting, or reply –

Scene xii

ARNULPH, ELBERTA.

ELBERTA. Scarce hadst thou left me in that loved retreat
Where five dear years I liv'd, and joy'd to live
For the short moments thou couldst kindly wrest
From clashing War, proud pomp.
He knew me for the daughter of King Ethelbert,
Who five full years, conceal'd, disguis'd, had 'scap'd
Offa's fierce search; but still he knew me not
For Arnulph's wife, the Mother of his Children! –
Think of my despair!
Torn from my Babes –
Acca, the faithful Acca, inform'd
Whither I hied – and werefore not return'd!
But the good Wilfrid, pitying my dire plight.
When I found
To my own castle they conducted me

Scene xii: El. 58 summarizes Scene xii as 'Relates her seizure and distresses. He his detention'. El. 46 contains a lengthy summary:

Arn‹ulph›'s First enquiry is for Children – She satisfies she left them safe: and[1] says Her Fathers disgraced and disbanded soldier and servant met her, spoiling to win his return in Arnulph's absence, and knowing, instantly seized her – and gaining help, conveyed her to Offa, as a prize for reward, being returned unpermitted from Banishment. Offa instantly offered her marriage. She durst not own her situation, fearing to involve Arnulph for marrying proscription – Her agony for her forsaken Children – Wilfrid, a faithful Servant.[2] undertook to get her sight of them. But he had not returned, though she had every hope, as he had nobly succeeded in telling her tale to Ceolric.

Expects Babes are on the road – Joint joy. Says he will watch for them himself. Doubts not the amnesty must release her: May then own marriage to the Conqueror[3] and claim her and Children, and fortune from Offa.

He relates the civil command which detained him from her.

[1] 'Elberta' is deleted.

[2] several lines are deleted:
of her Father, who lived with him to his death abroad, and returned but as she was provided with a protector in her husband, – forced by Offa to come to the Castle with accounts of rents, saw her from her window – knew her, continued to stay on pretence of business in the House till he could speak to her – heard her tale, and went on an embassy to bring her children, and poor deserted Acca, to [two illegible words] hard by as children of Acca – where she might see them on the distant heath from her high Chamber.

[3] an insertion, replacing 'Stephen'.

I xii 1 s.d.–4 *ARNULPH ... pomp*: from El. 48. The speaker's name is an editorial insertion.
I xii 5–8 *He knew ... his Children*: from El. 7. The speaker's name is an editorial insertion.
I xii 8 *the Mother ... Children*: an insertion, replacing 'nor knew the generous Arnulph'.
I xii 9–20 *Think of my despair ... shamed*: from El. 30. The speakers' names are editorial insertions.

My Father's –
Then how I dreaded the proud hate of Offa
O how – how far more dire I dread his love!
Still I rejected – but still durst not own
That wretch whom erst my Father execrated
Disbanded – shamed –
ARNULPH. O leave me!
He will be here.
ELBERTA. Deny me not this Moment!
Parted so long – so hardly met!
ARNULPH. I must be gone –
I dread the approach of Offa –
ELBERTA. Nay – a little moment –
Long may it be ere we may meet again!
A little moment –
ARNULPH. Think of the danger.
ELBERTA. No! – think of the joy!
ARNULPH. Hadst thou but waited –
ELBERTA. O had I waited – but another day
My little Eric had perchance forgot me!
O but to hold them to my Breast once more!
Think of the bliss to see them once again
To feel in sweet embrace their gentle Limbs,
Soft on their Cheek to print the tender Ki‹ss›
And view the opening smile of dawning memory
That cries Ah! is it you, my mother?
The soft entwining of their little arms
Th'unutterable enchantment of their smiles –

I xii 30 *Eric*: cf. Act III, Scene i, where he is called Edwin.

I xii 20–27 *O leave ... little moment*: from El. 8. The speakers' names are editorial insertions.
I xii 30 *little*: an insertion, replacing an illegible word.
I xii 33 *in sweet*: an insertion, replacing 'their soft'.
I xii 35 *And view*: 'To see' is given as an alternative.
I xii 35 *memory*: three lines are deleted:
That conscious recollects the early kindness
That cries Ah! is it you, my Mother?
Which first
I xii 27–38 *Think of the danger ... their smiles*: from El. 32. The speakers' names are editorial insertions.

Scene xiii

ELBERTA, ARNULPH.

Enter OFFA.

ELBERTA. All that is female in me now of fears
Confusion, terror, and timidity
I banish, to give place to firmer virtues.
Slander calls forth a courage new to me,
A vigour that disdains to shield itself
Beneath the battery of a trembling nerve
Seasons make all and calumny endur'd is deserv'd:
Our sexe's modesty becomes a traitor
To honour, liberal worth, and noble feelings
Where it will sink us 'neath the sable shelf
That robs us of good Name
Far from the praise retir'd of female modesty
Be the mute silence that submits to slander
Her fair unspotted chastity of fame.

Scene xiii: El. 58 summarizes Scene xiii as 'Surprises them. Rage. Accuses Elberta of treason'. El. 9 provides a longer summary:

Offa enters unexpectedly; starts – they are confounded. He challenges their business together.[1] He challenges their former acquaintance.[2] Arnulph says he may find a time to right a lady: she is silent.[3] Arnulph breaks forth in her defence: Offa says she is leagued with Malcolm to place Edgar Atheling on the throne,[4] and furious accuses her of treason. Arnulph fires.[5] Elberta takes courage, comes forth, and spiritedly defends herself.[6]

El. 31 states more briefly:

Arnulph boldly claims his wife supported by
When she is arrested for treason – and he is sent to the army on a Commission.

The dialogue is from El. 31.

[1] an illegible word and 'villain' are deleted.
[2] an illegible word and 'villain' are deleted.
[3] an insertion, replacing two illegible words.
[4] 'Offa says . . . throne' is an insertion, replacing 'Offa, furious, says he will take his' and an illegible word.
[5] two illegible words are deleted.
[6] 'Her Father died in a loyal cause, and blessed his Exit so dying to' is deleted.

Scene xiv

ELBERTA, ARNULPH, OFFA.

Enter SAXRED.

Scene xv

ARNULPH, OFFA (*manent*).

OFFA. And now, sir, what were your commands with me?
ARNULPH. The Peace
Summons each Chieftain to renew his Homage.

Scene xiv: there is no dialogue for Scene xiv. El. 58 summarizes it as 'He confirms Treason, affirms Proofs. She retires indigna‹nt›'. El. 39 provides a longer summary:

Offa enquires after suspicious papers. He avers that Informo picked them from the ground, falling from the robe of Elberta as she descended to walk. Fully containing proofs of a league for Edgar Atheling promised by her late Father, and to be executed by her and her friends. She exclaims – but refuses personal altercation, proudly, with such an antagonist, and committing her Cause to Arnulph and Truth, retires accompanied by Saxred.[1]

[1] 'accompanied by Saxred' is an insertion, replacing '[an illegible word] Saxred [an illegible word]'. Several lines are then deleted:

Arnulph haughtily tells Offa he may refer the matter to the King at Winchester[1] whither he has in order to command him on his fealty to pay new Homage and swear the succession to young Henry Plantagenet of Anjou.[2]

The incensed Offa orders him back to answer as an accomplice, under arrest.

Forced to withdraw, he assures Elberta of his protection.

[1] probably still the second largest city in England at the time and the seat of the treasury. According to Hume, Winchester was William's 'usual place of ... residence' (*History of England*, I, 222).

[2] Matilda's son; later Henry II (1154–89).

Scene xv: El. 58 summarizes Scene xv as 'Arnulph orders him to the King, menaces, and away'. El. 13 provides a longer summary:

Arnulph haughtily tells Offa he may refer the affair to the King, from whom he has it in charge to command him instantly to Winchester at a grand meeting of Peers and Chiefs upon the Peace with Scotland.[1] Then bids him on his peril offer insult or affright, meantime, to Elberta,[2] whom he shall return with higher authority to reclaim.

[1] 'at a grand ... Scotland' is an insertion, replacing 'to renew his Homage, and swear allegiance to the young Henry of Anjou'.

[2] 'and away' is deleted.

I xv 1 s.d. *ARNULPH, OFFA*: 'SAXRED' is deleted.

I xv 2 *Peace*: an insertion, replacing 'amnesty'.

I xv 1 s.d.–3 *ARNULPH ... Homage*: from El. 12. The speakers' names are editorial insertions.

I xv 3 *Homage*: two lines are deleted:

And swear allegiance to the appointed Heir
Young Henry Plantagenet of Anjou.

OFFA. I will appeal – present your condemnation –
ARNULPH. Make your appeal. Mine also will be heard.
An Englishman shall never brook injustice –
His spirit rises from th'oppressor's rod
And even when most he's curbed and vex'd and chaf'd
Like the fleet Courser who disdains controul
While, with fervent zeal, he burns to serve
The Monarch – patron of his rights, not spoiler,
Something still whispers to his honest Heart
Allegiance must be pair'd with Liberty.

Scene xvi

OFFA *solo*.

OFFA. This fellow who, on chance alone of recompense
Could burst the bond of trust –

I xv 9 *Courser*: large, powerful horse. Burney uses the word anachronistically in the sense of a swift horse or racer, a meaning first current in the seventeenth century.

I xv 10 *While*: before 'While', 'Ev'n' is deleted.
I xv 10 *fervent*: an insertion, replacing 'honest'.
I xv 13 *pair'd with*: 'fed by' is given as an alternative.
I xv 3–13 *I will appeal ... Liberty*: from El. 40. The speakers' names are editorial insertions.
Scene xvi: El. 58 summarizes Scene xvi as 'Resolves all perjuries, and to commit Elberta to Ceolric's Tower'. El. 41 provides a long summary:

Rage and despair about Arnulph's pretensions determined to carry all sooner than permit a marriage that demands [an illegible word] restitution. Civil War, misery to all from its vicissitudes! Even to those who have nothing precious to lose.

Forced to go, durst not leave Elberta lest Arnulph attempt her release: with this spy resolves to formally make her over to Ceolric, for his strong Tower, who, believing the tale or not, must equally guard her for further[1] orders.

Then, accompanied by Saxred and Informo, to fly and lay the treason before the Conqueror,[2] and get her confined for life, and his succession secured.

And, meanwhile, harrassed with his casket, to deliver it also to the keeping of Ceolric, whom, though he hates, he knows to be one who durst not be dishonest.

Whatever be the wishes of Ceolric, perchance, once in his Tower, she is a Prisoner of State, and cannot be removed but by order, under penalty of Treason.

Sure of some Spy – Distrusts every one – Horrible state!

The dialogue is from El. 35. El. 258, which is marked 'if wanted for 2d. act', dramatizes the transfer of the casket to Coelric:

OFFA. CEOLRIC.

OFFA. Take this Casket
Of value singular – pledge me thy oath
To guard it safe –

[1] an insertion, replacing 'her'.
[2] an insertion, replacing 'Stephen'.

May surely be persuaded
I dare not disappoint the treacherous wretch
Lest he in vengeance
If I trust Saxred
Some demon may seduce him to purloin
No, Ceolric's the man
His conscience is unhardened.
He will not dare
The smallest deviation from strict rule
Would startle him in dreams –
'Tis Ceolric
That will believe the matter to be true –
Why doubt? There's proof.
'Tis fortunate he's here.

ACT II

Scene i

CEOLRIC.

A Parade before a Tower.

CEOLRIC. She false? She trait'rous? She immersed in Plots?
I will believe it when this good right hand
Subscribe me her accomplice. Times are bad,
And Man is wicked: but so bad, so wicked
To give to deep hypocrisy and feint
The semblance of such virgin innocence. –
Mysanthropy itself
Nature and Instinct govern Man within –
Without . . . the motley habit which he wears
Is not more artificial than his conduct –

Scene i: El. 66, which provides a synopsis of each scene of Act II, describes Scene i as: 'Resolves to protect Elberta'. El. 74 reads: 'Disbelieves the assertions of Elberta's guilt, yet rejoices he has her in custody, that he may soften her confinement till an explanation brings enlargement'.
II i 1 s.d.–7 *CEOLRIC. A Parade . . . itself*: from El. 75.
II i 8–10 *Nature . . . conduct*: from El. 261.

Scene ii

CEOLRIC, SOLDIER.

Scene iii

CEOLRIC, ARNULPH *as stranger*.

ARNULPH. Five years since,
When we were yet allies, not Enemies,
I lov'd the beaut'ous princess – meant to claim her of Ethelbert
Wilt thou not suffer me to once behold her?
Wilt thou not grant to a despairing Husband
One short, quick, fleeting moment of adieu?
CEOLRIC. My life is in the charge.
ARNULPH. Mine, in thy will.
CEOLRIC. How camst thou free? which way? by what permission?
ARNULPH. The powerful Vortigern has broke the barriers –
Confusion in the army – loss – disorder
Have forced my stubborn Uncle to release me

II iii 9 *Vortigern*: Hume writes that Vortigern, Prince of Dumnonium, persuaded the Britons to invite the Saxons to bolster their defence against invading Picts and Scots in 449 (*History*, I, 18). Burney may also have remembered William Henry Ireland's forgery *Vortigern*, alleged to be a newly discovered play by Shakespeare, which was performed at the Drury Lane theatre, 2 April 1796, and was published by Ireland's father Samuel in 1799. Vortigern's presence in this play is another of Burney's anachronisms.

Scene ii: there is no dialogue for Scene ii. El. 66 summarizes Scene ii as 'A stranger claims admittance'. Similarly, El. 76 says 'A stranger desires admittance. Granted'. El. 77 adds 'A man of mean array, but urgent speech claims instant parley'.

Scene iii: El. 66 summarizes Scene iii as 'Begs interview with Elberta'. El. 78 provides a longer summary:

> Arnulph says the[1] arts of Offa have prevailed for the present, tho' truth must arise; and the King resolves to keep her captive till the witnesses are heard: he himself is ordered instantly to join the army.[2] He conjures to see his wife first. Ceolric blames his hasty marriage. He relates meeting her and her Father in Exile, and there seeing her in simplicity of [an illegible word] united himself to her – on her Father's death, she came over and here, had[3] lived retired five years, in disguise, – he visiting only by stealth and incognito. Fearing perjuries of Offa, durst not avow Children till her release and acquittal and dies to see before he goes.[4] Gives his word not to attempt her personal rescue; and is admitted.

[1] 'valiant Envoy found his secret retreat from army' is deleted.

[2] an insertion, replacing 'ordered to the continent, attending an Embassy – '.

[3] 'meeting her . . . over and here, had' is an insertion, replacing 'its [an illegible word] he took her prisoner after already loving her. Ceolric says he should have married her also so circumstanced. He married and they'.

[4] 'Envoy had aided his secret absence. He had confided in his friendship to the Father's Memory' is deleted.

II iii 1 s.d.–11 *CEOLRIC, ARNULPH . . . release me*: from El. 88.

To head once more his army to the West.
No more against East Anglia will he trust me –
The soldiers think me
The valiant Eric aided my empire,
And, thus disguis'd – hither I sped my way
To beg one last dear word with my Elberta.
Five years since –
When first this arm knew victory – and routed
The forces of East Anglia – all combin'd
Beneath brave Ethelric, to call down vengeance
On Mercia and on Offa for the slaughter
Of Ethelbert – base and inglorious deed! –
'Twas then, an hapless fugitive, his Daughter
Whom all believed destroyed – the fair Elberta –
Ta'en by a common soldier, in her flight
Unknown – all trembling – to my tent was brought –
Ah! when shall I forget
With youthful terror – supplicating beauty –
Low at my feet she fell – I rais'd – I knew her –
Quick re-assur'd her courage – pledg'd my Honour
To guard her from her Father's murderer – Offa, –
And, in a lone, but [1 illegible word] village lodg'd her –
There, all conceal'd, she dwelt – there, sacred bonds
Made her my own – there with two lovely Babes.

CEOLRIC. Is't possible –
ARNULPH. Most true.
CEOLRIC. Two children.
ARNULPH. Ah hush! The sound alarms – O keep it sacred –
Should the stern Offa learn
There breathes a little Heir to either Kingdom,
To Mercia – and East Anglia
How might my lonely Boy escape his rage?
Who should secure his little Sister's safety?
O should he know the Grandson of that Ethelbert
He slew in cold deliberate treachery . . .
But Ethelric, brave General of East Anglia,
Advances fast –

II iii 12 *once more*: an insertion, replacing 'the Kentish'.
II iii 12–26 *To head . . . in her flight*: from El. 88 verso.
II iii 27–35 *Unknown . . . lovely Babes*: from El. 91.
II iii 36 *ARNULPH*: 'Two Child.' is deleted.
II iii 36–46 *CEOLRIC. Is't possible . . . Advances fast*: from El. 79.

Scene iv

ELBERTA *sola.*

Scene v

ELBERTA, CEOLRIC, ARNULPH *disguised.*

ELBERTA. O Arnulph, my lov'd Arnulph
Where are our Children? What may now betide them
Who may receive their
CEOLRIC. There, still unknown, thy Children might remain
Yet there I'll place the Babes – there see and kiss them –
All I could raise, or part with unsuspected
Is gone this morning to their secret use. –
ELBERTA. No matter – save them but – and leave all else
To better times – or worse!
CEOLRIC. Times I defy
When thou and they are safe.

Scene iv: there is no dialogue for Scene iv. El. 66 provides a one-word summary, 'Despondence', while El. 80 adds 'despondence – fears for Children'.

Scene v: El. 66 provides a short summary of Scene v: 'Plan protecting Babes by Ceolric'. El. 82 provides a lengthy summary:

Elberta expresses her agony for her Children. Asks if no news of Wilfrid? Says he may go to castle and get no information, by Offa's acts, of her confinement. Children must be at the Cottage on the Moor. Arnulph would go to Cottage: but duty and necessity[1] call him back to his post.[2] She supplicates Ceolric to let them know her fate, and give them courage. He yields. Promises to give money and directions to Acca to stay[3] with them there till further orders. But having promised to deliver a casket to Fortress for Offa – must defer till morning. Arnulph entreats that commission, which he can execute in his way back, unknown. It was only to be delivered, with a paper[4] from Offa, to a man of trust, there resident.

He consents[5] to leave orders Wilfrid may see her if he comes.

[1] an insertion, replacing 'honour'.
[2] 'He could by aid from Envoy. That would be too late' is deleted.
[3] an insertion, replacing an illegible word.
[4] an insertion, replacing 'note'.
[5] 'They separate' is deleted.

II v 2 *our*: an insertion, replacing 'my'.
II v 2 *What ... betide them*: an insertion, replacing 'Who shall' and an incomplete word.
II v 1 s.d.–3 *ELBERTA, CEOLRIC ... receive their*: from El. 84. The speaker's name is an editorial insertion.
II v 4–10 *There, still ... they are safe*: from El. 85. The speakers' names are editorial insertions.

ELBERTA. There they may wait, and wander, and expire –
O were they here! . . .
Their little Hands, uplifted, should implore –
Should
With eloquence that distances all words –
The very picture distances all words –
O pity! pity! generous Ceolric!
CEOLRIC. By Heav'n an if I had an Heart
Hewn out of Scyllas Rock
I could not but give Ear!
I'll save them!
ELBERTA. O most noble!
CEOLRIC. Yes! – myself! –
And – you shall kiss them still –
ELBERTA. Sweet! sweet and good! – I cannot thank
My o'er charg'd Heart –
Their little lips alone can.
ARNULPH. In this plain simple guise, none will suspect
I have ought to guard –
And fair Simplicity – Daughter of Truth
Of innocence, of modesty, of candour.

Scene vi

WILFRID.

A walk before the Great Gate of the Castle.

WILFRID. This is the heaviest news that ere I heard.

II v 19 *Scyllas Rock*: a rock upon the Italian side of the straits of Messina, personified as a dangerous sea-monster with twelve feet and six heads, each with three rows of teeth.

II v 17 *generous Ceolric*: 'Elb' is deleted.
II v 11–25 *There they may ... lips alone can*: from El. 86. The speakers' names are editorial insertions.
II v 29 *Of*: 'lovely' is deleted.
II v 26–29 *In this plain ... of candour*: from El. 83. The speaker's name is an editorial insertion.
Scene vi: El. 66 summarizes Scene vi as 'Heavy grief at Elberta's seizure'. El. 67 provides a fuller summary:

> He comes from the Castle wringing his hands, and lamenting this is the heaviest Day he has known. Too fatigued and weak to proceed, he seats himself beside the Castle Gate, in a desponding posture. Repines at illness that made him too late at last to see Elberta before her captivity. Certain of her innocence, and unlawful accusation.

The dialogue is from El. 68.

Had my dear Lord yet liv'd, this would have gall'd him
Beyond each sore affliction personal.

Scene vii

WILFRID, ENVOY , INFORMO.

Scene viii

INFORMO.

Scene vii: there is no dialogue for Scene vii. El. 66 summarizes the scene as 'Wilfrid and Envoy form a compact to secure Babes'. El. 69 provides a much longer summary:

Wilfrid, seated, enter Envoy, also from the Castle, attended by Informo.

He raves at missing Elberta; says he must see her whither soever she may be conveyed. Informo says her guilt must first be attoned. Envoy vows he disbelieves it – Wilfrid suddenly rises, and laying both his Hands on his arms, passionately thanks him, protesting he knows her to be innocent as the Babe unborn. Envoy enquires; hears 'tis his friend's old Servant; embraces him for his fidelity.

Envoy tells Informo to shut the Gate – and Wilfrid acknowledges[1] the marriage and children.[2] Speaks of Babes – and owns they are at the Cottage on the Moor. He desires him to return and guard them safely and in secret,[3] with Acca,[4] as her Children, promising them his entire protection till their Parents can claim them, from love to Ethelbert's memory, and detestation of Offa. Meanwhile, he undertakes to procure the release of Elberta from the King, or admission to see and counsel her. Ceolric he knows to be one of the bravest and most humane of men, and is sure she will be well treated – he will to Court, to plead her Cause, and curry her a second release.

[1] 'to Envoy' is deleted.

[2] 'Envoy was he says [an illegible word] which he had heard from Arnulph at Winchester. Wilfrid' is deleted.

[3] an insertion, replacing 'to his own Seat'.

[4] 'still' is deleted.

Scene viii: there is no dialogue for Scene viii. El. 66 summarizes the scene as 'Overheard, and will pursue'. El. 70 provides a slightly longer summary: 'Congratulates himself he has overheard the chief of their discourse, enough to pursue to Cottage on the Moor'.[1]

[1] 'Resents not having casket, or knowing where lodged' is deleted.

Scene ix

INFORMO, SAXRED.

Scene x

SAXRED *solo*.

SAXRED. Pretends he now to secresy with me?
Why what becomes of him should I inform.

Scene ix: there is no dialogue for Scene ix. El. 66 summarizes the scene as 'Says he will follow and spie'. El. 71 provides a lengthy summary:

Informo tells Saxred he waited but his return to inform him he was off – and to follow[1] Offa, and tell him[2] that he was pursuing old Wilfrid, who had some deep scheme with Envoy, which he would fathom, and let him know the result.

Resents not being trusted with Casket, asks if Saxred has it? No; neither know of it.

Informo goes off grumbling, yet full of business and plan, and hope of profit.

Bids him let Offa know his route, and where he will watch to receive orders.

Says there are two children at the Cottage on the Moor – Saxred[3] says they must inherit, – deepest importance to acquaint Offa with the intelligence. Informo appoints a spot where he can receive instructions what to do. Meanwhile, will watch, and if they quit Cottage, pursue.

[1] an insertion, replacing 'tell'.

[2] 'and tell him' is an insertion, replacing 'should he return'.

[3] 'doubts' is deleted.

Scene x: El. 66 summarizes Scene x as 'Resolves to search Profit, and turn honest'. El. 73 adds a longer summary:

No faith or confidence in Offa – no pleasure or harmony. Longs to be emancipated from serving him. Wishes he could be found out without involving himself for vengeance.

If there are Children doubts whether 'twill be worth while to abet Offa, who can then[1] no more possess for success than for failure – yet – so involved –

The dialogue is from El. 72.

[1] 'niether' is deleted.

II x 1 s.d. *SAXRED*: before 'SAXRED', 'Informo' is deleted.

ACT III

Scene i

COTTAGER, ROWENA, INFORMO.

A Cottage.

Scene ii

COTTAGER, ROWENA, INFORMO, ARNULPH *disguised.*

ARNULPH. My Elberta!
Couldst thou have shared the blessing of this moment.
Who brought them hither? – say, what gentle Hand
'Tis then my noble Ceolric
O Heaven! sweet Heav'n! show'r blessings on his head.

Scene i: there is no dialogue for Scene i, but El. 93 provides a summary:
Informo sifts artfully who they have at the House; hears only two Children, and a Female who has the care of them. He asks if they have been sought by any one. Yes – this very night[1] a Gentleman brought them money. Who? They know not, but he was very rich, for he gave them much,[2] only to let them stay till he sent or came again. He was so muffled in the Face, they should not know him if he came again. The woman delights in the Children. The men, in the reward. Informo begs to stay the night. They consent. A knocking. Suppose some Traveller benighted –
[1] 'have been sought ... night' is an insertion, replacing 'came by themselves? No'.
[2] 'money' is deleted.
Scene ii: El. 94 provides a brief summary of Scene ii:
Arnulph anxiously enquires for Children. Hears they are safe – trembles to see them – represses – asks with whom? a female guard. Informo, starting, muffles up, and walks aloof. He cannot resist – and visits them. Returns enraptur'd.
III ii 1 s.d.–2 *COTTAGER, ROWENA ... this moment*: from El. 94.
III ii 3–5 *Who brought ... on his head*: from El. 95.

Scene iii

COTTAGER, ROWENA, INFORMO *walking aloof*, ARNULPH.

Enter ACCA.

ARNULPH. When there – abide there –
Ethelric will receive – will
COTTAGER. How know I that? I have no word but yours
And you? Who are you?
ARNULPH. If – at once – in hand – in sight –
This same reward –
This villain, this mercenary wretch
Will serve me but for gain – yet –
Why do I call him mercenary villain
What claim have I upon his risk? his service?
Have I for him done ought? out! haughty Egotism.
Because already I possess the most
Must I of him who never knew my blessings
Demand . . .

Scene iii: El. 96 provides a long summary of Scene iii:
Acca comes to speak to him apart as he is going,[1] asking directions: he answers, as repeating, they will soon be protected by Ceolric, and must have patience – and[2] demands what is become of Wilfrid? He saw them, slowly, to the appointed place, and left them to inform Elberta of their safety and sejour; They had not seen him since; but a Gentleman had come, and told News from her, and[3] bid them wait with hope and courage. Yes! he cries 'tis the generous Ceolric. Acca hurries.
Arnulph desires Cottager to take a note for him to Envoy. –
He demands pay.[4]
Arnulph hesitates, and Acca says he had better be paid there, to avoid suspicion, and make him pass for a poor man –
The man refuses.
He raves – then pardons – and he is turning[5] to Informo who walks off, when a knocking alarms –
Dreads being known.
Tells Acca he comes on furlough – stops opening door –
The knocking more violent.
Acca and Arnulph retire to Children's room.
[1] 'as he is going' is an insertion, replacing 'before'.
[2] 'protected by . . . patience – and' is an insertion, replacing 'fetched to Envoy'.
[3] 'conducted them to this Cot' is deleted.
[4] 'first' is deleted.
[5] 'he is turning' is an insertion, replacing 'is applying'.
III iii 1 s.d. *ARNULPH*: 'COTTAGER's SON' and 'SPY' are deleted.
III iii 1 s.d.–4 *COTTAGER . . . Who are you*: from El. 97. The speakers' names are editorial insertions.
III iii 8 *gain – yet*: 'hold' is deleted.
III iii 5–21 *If – at once . . . render precious*: from El. 98.

Should he of me demand some active service
Some toil to take but one poor Day from life
In his behalf – should I not stare at him
As if Insanity or insolence
Projected the strange wish?
And why? because to me he cannot offer
That which he has not! Nor can render precious
That which I lack not?
O, harried from reflection, what is man?
How wide from Reason! how estrang'd from Justice.

Scene iv

COTTAGER, ROWENA, INFORMO.

Enter Wilfrid *and* Elberta.

WILFRID. Is this your hospitality to Travellers
Benighted and distressed?
ELBERTA. Waste not an instant now in vain reproach
Good Cottagers –
Say if two Children, lovelier than the morn
Sweeter than

Scene v

COTTAGER, ROWENA, INFORMO, ELBERTA, WILFRID.

Enter ARNULPH.

ELBERTA. Live then? – my little ones?
Tell me but if my lovely children live?

III iii 22–24 *That which ... from Justice*: from El. 99.
Scene iv: El. 100 provides a brief summary: 'Wilfrid reproves for being kept – Elberta interrupts, to ask for Children'. The dialogue is from El. 101. The speakers' names are editorial insertions.
Scene v: El. 102 provides a brief summary: 'Arnulph rushes in in a transport of surprise and joy. Elberta, embracing, calls for her Children – and rushes eagerly in to them'. The dialogue is from El. 103. The speaker's name is an editorial insertion.

Scene vi

COTTAGER, ROWENA, INFORMO, WILFRID, ARNULPH.

Scene vii

COTTAGER, ROWENA, INFORMO, WILFRID, ARNULPH, ELBERTA.

ELBERTA. They are both at rest.
Lovely they look in Infant innocence
And with the sleep of angels seem Drest.
How roseate on their Cheeks glows blooming Health!
How beauteous o'er their forms spreads native grace!
How sweet they look! Safety sits smiling o'er them.
O, I have lost all terror since I have seen them!
What dimpling cheeks! My Arnulph, I could trust them

Scene vi: there is no dialogue for Scene vi. El. 104 provides a summary:

Arnulph makes 100 enquiries, and hears briefly Wilfrid had missed Elberta at Castle, and met with Envoy, and his promised succour, and support of Children. That, arriving at the Tower, he was admitted without difficulty to Elberta, and found all so slightly watched, that he planned with ease, from the absence of Ceolric, her escape, that she might herself come to see her Children.

Arnulph is dissatisfied for Ceolric –

Scene vii: El. 105 provides a full summary:

She comes back in raptures. He blames her flying Ceolric. She justifies by Edwin's Illness; and next by fears of Offa's acts prevailing.[1]

He has obtained a second furlough – but durst not be known.

They both fear Offa's hearing of their Babes.

Informo listens insidiously.

He purposed but to deliver Casket, – but could not resist using his furlough and freedom to take one view of his babes. He now fears being too late to go round to Fortress, and gives Casket to Wilfrid to deliver.[2]

Horror lest Ceolric should be blamed.

She offers to return to him, having seen Babes and Edwin well –[3]

Dread Offa should hear of two young Heirs.

Informo approaches. Arnulph alarmed, asks who he is.

Elberta recognizes him.

He threats. They are forced to bribe.

He goes off with extorting all they possess not to impeach Children.

[1] 'Deep distress what to do' is deleted.

[2] several lines are deleted:

Knows not where to leave her – how to part – She offers to return to Ceolric but he dreads Offa and Wilfrid thinks once in his power, he will suborn witnesses to destroy her.

[3] 'Conflicts cruel' is deleted.

III vii 8 *dimpling Cheeks*: 'cherub sleep' is given as an alternative.

III vii 9 *ravenous*: an insertion, replacing 'Tygers'.

III vii 1 s.d.–11 COTTAGER ... *instinctive pity*: from El. 106. The speaker's name is an editorial insertion.

Fearless, midst ravenous wolves, and hungry Vultures,
Their unresisting helplessness would move
Even Beasts of prey to soft, instinctive pity.
Thy uncle I could brave, with all his hardness,
Thy soldiers I would meet with all their arrows
So in my arms I could but hold those Innocents
And bid them, ere they struck – but once, survey them!
Blest prejudice! by Nature's kindliest Laws
Lent to sustain maternal tenderness!
Without such fond credulity, what peace
In such a world of wrongs, could sooth a mother?

ARNULPH. Elberta, from a strength of mind like yours,
So little spoken by its sexe's weaknesses
So open, noble,
Could I expect this? Frivolous impatience
Should have no weight in Honour's scale
And where is honour binding, if not here?

ELBERTA. He told me our sweet little one was ill –
Ill, – suffering – and yet motherless! – O Arnulph –
So young, so tender, –

ARNULPH. Ill? – Edwin ill? –

ELBERTA. Yes – Dost thou blame me, Arnulph? –
Was the decision easy? conflict light?

ARNULPH. I am glad I was not there!

ELBERTA. Ah, Arnulph! – See I not thy sympathy
Yes! all the Father flushed in thy face!
Honour is noble . . . but
Is their a tie superior to paternal?

ARNULPH. Honour – the first tie of manly responsibility.

ELBERTA. No Arnulph – Honour is indeed most noble
But there something is of yet higher Cast –
Innocence!

ARNULPH. My Honour – O Elberta
My Honour thus assaulted unrevenged –
My Honour seems to me already tainted –

III vii 13 *soldiers*: an insertion, replacing 'army'.
III vii 14 *in my arms*: an insertion, replacing 'to my Breast'.
III vii 17 *tenderness*: 'What peace could' is deleted.
III vii 12–19 *Thy uncle . . . sooth a mother*: from El. 106 verso. The speaker's name is an editorial insertion.
III vii 30 *Was the*: before 'Was', an illegible word is deleted.
III vii 33 *Yes*: an insertion, replacing 'Aye'.
III vii 20–35 *Elberta, from a strength . . . to paternal*: from El. 108. The speakers' names are editorial insertions.
III vii 36 *Honour – the first*: before 'Honour', 'than love' is deleted.
III vii 38 *there something is*: an insertion, replacing 'is not Innocence'.
III vii 36–41 *Honour – the first . . . already tainted*: from El. 244. The speakers' names are editorial insertions.

ELBERTA. If Saxred knew
That here the little Heir of Ethelbert
The lawful Heir.
ARNULPH. Who is that man, who sitting there aloof
Heav'ns! is it . . . Uddolph?
ELBERTA. Is it then Uddolph?
ARNULPH. Yes –
ELBERTA. O blight to every hope!
Gain is his Idol –
INFORMO. How are you, my good General?
ARNULPH. Ha! Eric –
Thou base deserter! thou accursed Spy?
INFORMO. Am I so naught! – And in your Eyes, too, General?
ARNULPH. Avaunt! –
INFORMO. Give me then something to support Existence.
ARNULPH. Yes, this I give thee –
ELBERTA. Hold hold!
Cruel is the death thy rage, tho' just, inflicts –
ARNULPH. What? – reward such baseness –
ELBERTA. Think of these little ones! betray'd –
ARNULPH. Fellow! – take my pursc –
And heavy on thee fall its vile abuse!
INFORMO. And is this all?
Nay, I must yet have more –
Had I told Saxred he had giv'n me more –
O delirious wealth!
How bright it shines!
This is a facile way to make a fortune –
To fright poor fools –
If I should now tell Saxred I might double it –

III vii 46 *Uddolph*: i.e., Informo.
III vii 52 *avaunt*: go away, depart.

III vii 42–44 *If Saxred . . . lawful Heir*: from El. 107. The speaker's name is an editorial insertion.
III vii 45–46 *Who is . . . Uddolph?*: from El. 109. The speaker's name is an editorial insertion.
III vii 46–48 *Is it then . . . Idol*: from El. 110. The speakers' names are editorial insertions.
III vii 49–61 *How are you . . . yet have more*: from El. 111. The speakers' names are editorial insertions.
III vii 62–67 *Had I told . . . double it*: from El. 112. The speaker's name is an editorial insertion.

Scene viii

COTTAGERS, ROWENA, WILFRID, ARNULPH, ELBERTA.

ARNULPH. O what a curse
To see triumphant villainy! to know
Where guilt resides, yet dare not punish it!

Scene viii: El. 113 provides a detailed summary:
Arnulph's rage at Informo's success.
Deep distress what course to take.
Fears this wretch may play false doubly –
Terror for Children once known to Offa.
Desire to send them to King Malcolm till the turn of affairs be known, and think Elberta herself can there alone be safe, till[1]
But how? no money remaining.
Elberta stript in Castle.
Wilfrid and Acca have nothing – but what must pay Cottagers.
Wilfrid proposes Casket.
He resents.
Says 'tis probably Elberta's Wealth –
They know not that – and 'twas in Ceolric's trust, which renders it sacred –
Yet they must haste away – Ceolric himself must pursue – and to be *taken*, not [an illegible word]! –
Wilfrid knows a Cave peculiarly formed where they may wait till Arnulph consult Envoy – or Ceolric.
They part for this.
Arnulph from Door, – Elberta, and Cottagers, to Children
King Malcolm receives well all the women friends and adherents of Edgar Atheling –
El. 125 develops the idea of the cave:
An artificial excavation is described by Wilfrid,[2] dug in remote times, and for secret residence or repository, by a rebel Baron – where they may be safe till they can proceed elsewhere and Arnulph can advise with Envoy. A fit of delicacy keeps them from Ceolric.[3]
[1] 'Desire to . . . safe, till' is an insertion, replacing:
Desire to place them with Envoy.
How? no money remaining.
Elberta stript in prison Castle.
Desire and the King is acquainted with their birth and claims. Think Envoy might be overpowered, but to Envoy because not acknowledged.
[2] an insertion, replacing 'Ceolric'.
[3] several lines are deleted:
Offa, upon the amnesty published, gives a casket of richest treasure to Ceolric to convey, unsuspected, to a Fortress to prevent its claim.
To confine Elberta, invents forgeries of papers etc.
III viii 1 s.d.–3 *COTTAGERS . . . punish it*: from El. 115. The speaker's name is an editorial insertion.

Short is thy triumph, villainy!
Wait but for Night and solitude – 'tis o'er.
How can distress unknown await relief?
Durst I proclaim –
Could I but safely and in secret place thee
Where thy fair blaze of beauty were unnoticed –
Where our lov'd little ones might wile the Hours
Of absence, doubt, and fear – Then should I go
So buoy'd with hope, by such fond views supported,
That victory would hang upon my arm
And lead me dauntless through an Host of foes.
But – if dejected and forlorn I quit thee,
Thy miseries will bow down my fiercest courage,
Thy pallid face will be my nightly spectre,
And shrieks of Woe, or famine, – strike my Ears
Midst shouting multitudes that call to Battle
What! leave thee thus?
All unprovided ev'n with daily food?
No known resource? no bidden entertainment.
But day from day to hang on accident,
On Charity? on chance? –
Sooner would I abandon Honour's post
Sooner risk Death – nay, more, disgrace than bear it.
WILFRID. Sir – thou knowst the booty –
ARNULPH. Ha! – speak not of it!
WILFRID. Yet to starve – to perish –
ARNULPH. Hist, Wilfrid hist! try not my tremulous faith
With thoughts so perilous.
WILFRID. Was't not thy prowess obtain'd it?
ARNULPH. Hist! hist! Reason not, I pray.
Where Reason ballances the trembling scales
That poise temptation against suffering
Trust not her arguments! – far safer yield
To instant impulse, – which by feeling sway'd,

III viii 29 *Hist*: be silent.

III viii 4–5 *Short ... o'er*: from El. 114. The speaker's name is an editorial insertion.
III viii 6–14 *How can distress ... Host of foes*: from El. 116. The speaker's name is an editorial insertion.
III viii 14 *foes*: 'But if I quit thee, dejected and forlorn' is deleted.
III viii 15–19 *But – if dejected ... call to Battle*: from El. 116 verso. The speaker's name is an editorial insertion.
III viii 25 *Sooner*: 'As soon' is given as an alternative.
III viii 20–26 *What! leave thee ... bear it*: from El. 117. The speaker's name is an editorial insertion.
III viii 35 *arguments*: 'sophistry' is given as an alternative.
III viii 27–37 *Sir – thou knowst ... yet unerring*: from El. 118. The speakers' names are editorial insertions.

Impels to Right, unconscious, yet unerring.
To seize the treasure of another
Because or I or mine want succour?
Why where's the Highwayman – if this be reasoning
That reasons not as well – and estimates
Stern probity as justly.
If Honour on occasion rest,
If it be held on ground so slight and Casual
It may give way to incident – Why then
How differs he who
From he who thoughtless drags the plough
Uncultur'd in all nobleness of sentiment,
This Day, all kindness, aiding whom he loves,
The next, distemper'd, yielding all to self –
Since the vile perjury that thus pursues thee
Talks of the state . . . our noble Ceolric
Is bound to search – and should he find, secure thee . . .
Therefore – I dare not leave him the insignia
Where we may be addressed.

WILFRID. Not far from hence
Within the forest fair of Endermay
Hard by the borders of East Anglia.

III viii 56 *Endermay*: not further traced; probably fictitious.

III viii 38–42 *To seize . . . probity as justly*: from El. 119. The speaker's name is an editorial insertion.
III viii 43–50 *If Honour . . . all to self*: from El. 120.
III viii 53 *find*: 'to' is deleted.
III viii 51–55 *Since the vile . . . addressed*: from El. 121.
III viii 55–57 *Not far . . . East Anglia*: from El. 122.
III viii 57 *East Anglia*: El. 123–24 offers a possibility not mentioned in the summary. The name of Arnulph's friend is hard to decipher, both here and in the list of characters in El. 55, and is given conjecturally. He is described in El. 55 as a 'General and Friend of Arnulph'. 'With him . . . spirits calm' is from El. 124. 'Serene our minds . . . extremest roar' is from El. 124 verso. The rest is from El. 123.

ARNULPH. With him? no, no! I must not leave thee there
I know A‹shing›'s worth, his faith, his friendship: –
But the base world, all-judging by itself,
Might raise surmises to destroy our peace.
Elberta no! that peace, which not the goods,
The joys, the smiles of fortune know t'impart,
That peace which, midst affliction, stills the mind,
Which[1] parts calamity from turbulence,
Misfortune from reproach, and woe from phrenzy,
That let us guard, whate'er beside we lose.
ELBERTA. Yes, my lov'd Arnulph, yes; that peace of soul
Shall steer us still through the tempestuous blast,
And, midst its whirlwind, keep our spirits calm,
Serene our minds, compos'd our faculties,
And teach us, with the fostering gale of hope
Patient to stem the Storm's extremest roar.

Scene ix

A view of the Cottage at a distance.

Enter OFFA.

Scene x

OFFA, INFORMO.

OFFA. Bring me them quick – alive – or . . . even dead –
For . . . I have heard – fatigue is wond'rous speedy

No, let us wander – hapless – harrass'd – worn –
Dragging with pain our weary Limbs along,
Trembling with step infirm, shaking with fear
Ev'n at the zephyr's breath that wafts us on –
Sole febrifuge allow'd our fainting labour; –
Yet, in integrity, sure-footed,[2] firm,
Impervious there, to danger, hazard, toil,
Fatigue, distress – and more than all, Temptation.

[1] an insertion, replacing 'And'.

[2] an insertion, replacing 'firm-footed'.

Scene ix: there is no dialogue for Scene ix. El. 126 states merely: 'He thinks, by the direction of Saxred, that must be the spot'.

Scene x: El. 127–28 provide a lengthy summary:

He[1] has seen Saxred: and is just rode from his Company to hear news from Informo. Finding Elberta escaped he storms – and hearing of two Children, raves – and Arnulph – he will ruin – Bids Informo watch their proceedings, and dodge them from the Cottage to their retreat – and procure him if possible the Children. He will take care of them – their Memories are short, and they can tell no tales. As to Arnulph and Elberta, the Court is already so prepared, and Saxred so adroit, he fears not conquest. Encourages Informo to every hope: and bids him be vigilant, as all may be vanquished but young Heirs, who if once they escape them, render all else abortive, by seizing the whole revenue. Promises not to murder Children, at the request of Informo – but to secure them from interference. Tells him by these two Children he loses all. He has inserted Arnulph as an accomplice.

Informo says without previous pay for absconding he will not seize Children, as he is known even to Acca, and cannot therefore re-appear. Offa says he has a Casket of rich Jewels, and money at a fortress, for which he will instantly give him a demand at[2] sight of children – and with which he may be luxurious elsewhere for life. Says he will[3] shew him the spot where the Children must remain till dead of Night, then to be moved to a seaport, and shipped safely for other realms, as he will instruct him –

Bids him go round to meet him, that they may not accidentally be overheard, or observed together.

'He has seen Saxred . . . as an accomplice' is from El. 127. The rest is from El. 128.

[1] 'before 'He', 'Offa' is deleted.

[2] 'He insists on it immediately. Offa reluctantly complies. Gives it upon mention of opportunity' is deleted.

[3] 'Says he will' is an insertion, replacing 'Takes him apart to'.

III x 1 s.d.–4 *OFFA . . . young Beings*: from El. 129. The speaker's name is an editorial insertion.

Sometimes to cut the little thread of life
Of such young Beings –
So young –
Why what to them is existence?
What know they, at that age, of life's enjoyment?
I will not hurt them – O no! –
Yes! they shall live – though not, perchance, so
With all the luxury...
Why what of that? is luxury necessity?
May not a man be happy who
Bring me them hither....
I'll have the Children – that they live or die
Is of small moment –
Two Children said ye! Infamous indeed –
Let me behold them –
Think not I mean to – wrong – or injure them –
No – I would merely –
I pray you is it known that they have Children?

INFORMO. No.

OFFA. What! not suspected?

INFORMO. No. They have passed for Acca's since her captivity –

OFFA. Perchance they are not Elberta's in good faith.

INFORMO. O yes! I heard them say ...

OFFA. Out, fool! I say perchance they are not hers –
Dost thou not understand?

INFORMO. O ... why ...

OFFA. Such frauds are common where Estates impend –

Scene xi

OFFA.

OFFA. This fellow must away when all is o'er –

III x 5–19 *So young ... I would merely*: from El. 130. The speakers' names are editorial insertions.

III x 20–27 *I pray you ... Estates impend*: from El. 131. The speakers' names are editorial insertions.

Scene xi: El. 132 provides a summary:

Says he must manage this fellow a while, and then get rid both of him and Saxred, as they torment him to pay the troubles of their consciences.

He must away to Castle, not to be seen in this vicinity, his situation being suspicious.[1]

He will go first for Casket, which is too rich for such a Ruffian, and must be lightened.

Says he must escape him, to get round to the Fortress[2] unsuspected for arrangement.

[1] 'He has' is deleted.

[2] 'round to the Fortress' is an insertion, replacing 'the possession'.

III xi 1 s.d.–2 *OFFA ... his cowardice*: from El. 133.

Not for his virtue – but his cowardice –
That wretch pretends, at times, such mawkish qualms
As if not destitute of tender feelings –
Out on such hypocrites! a little pay
A bribery of some luxury or regale
Makes all forgot –
Yet in those lower ranks mean Superstition
Affects the airs of conscience, to impeach
Conduct and Sentiment. It fancies punishments
It forms vagaries of imagined warnings,
Sees supernatural sights – hears Screams i'th'air,
Is visited by menace of ill luck
And deems prosperity flies even success –
For me – I view with scorn deserved such folly,
Such feeble fear of unfix'd chastisements –
Could my firm mind but trust in other Men
As in my self – I were indeed invulnerable
But there the mischief lies! – I dread mankind
At large and individually alike!
I know no comfort, no sweet babes greet
While weak companions in my enterprize
Have power to blast me!

Scene xii

Cottager, Rowena, Ceolric.

The Cottage.

Ceolric. Gone? and together? –

III xi 2 *cowardice* – : Followed by additional summary: 'Says he must retreat, and on no account be found in vicinity of Children himself, his situation being suspicious'.
III xi 3–23 *That wretch ... to blast me*: from El. 134.
Scene xii: El. 135 provides a summary:

Ceolric, coming from the inner Room, repeats *Gone?* with rage and amaze – hears all together – he is searching Elberta, a state Prisoner, for whom he is responsable. Hears Arnulph by description of party – curses them, and dooms them to perish.

Hears their sadness, their tender meeting, and their soft Joy, and all appeased, vows to mislead his people, and suffer them to escape and be happy.

The dialogue is from El. 137–38. El. 136 contains an alternative version of the scene:

Ceolric. Gone?
All met?
Nay then – I'll every way pursue,
Nor rest till I succeed –
Ingrates! –
Cottager. Sad they all seem'd – yet – strange to say! all happy –
And tender, kind and loving ...

Let them all perish!
I will inform – pursue –
COTTAGER. They but expect it – wretchedness like theirs –
CEOLRIC. How? were they wretched?
Nay – what have I do to with pitying them
They merit it – away – Yes, let them perish!

going.

COTTAGER. And yet – amidst their woe – they seemed so happy! –
Clung to each other – smiled so through their tears.
Wrapt in their arms their tender little ones,
And said – the world contained no love nor joy
Like theirs – so they could live – or die together.
CEOLRIC. How – were they happy? –
Nay then – let them go!
And with them go my wishes, pray'rs, and blessings!
If happiness has dawned upon their sorrow
I cannot blight it. Safe let them go!
I will mislead my troop to aid their safe‹ty.›
Can I press them
To seize the unhappy fugitives? – divide
Ties so endearing? – Husband – Children – Wife –

CEOLRIC. Were they all here?
What, Husband – Wife, and Babes?
COTTAGER. Yes – and all fondling, though in Tears, and smiling, etc.
CEOLRIC. Let them then go! – I break not such a union
Let every risk be mine – save that of cruelty
O Sacred still be conjugal
Paternal tenderness be their's
And all the joys of soft domestic peace.
III xii 1 s.d.–18 *COTTAGER ... aid their safe‹ty›*: from El. 137. The speakers' names are editorial insertions.
III xii 19–21 *Can I press ... Wife*: from El. 138.

ACT IV

Scene i

ELBERTA.

Mouth of a Cave.

Scene ii

ELBERTA, ARNULPH.

Scene i: there is no dialogue for Scene i. El. 240 provides a short summary:

She comes forth from Cave seeking Wilfrid. Her Children want food. He is gone for it – and comes not – Inquietude for Arnulph, and if his return was in time – with him and her Children she could be happy in this desart place, if but free from alarm, and blest with competence for a modest existence.

Scene ii: there is no dialogue for Scene ii. El. 242 provides a summary:

He comes to her sadly – she flies to meet him – he enquires for Children – Wilfrid is gone to search them food, Wilfrid who has maintained them, with Acca! – She begs his history, and why he seems so gloomyly depressed? He answers, he has been arraigned, by vile Offa, as an accomplice, and his Honour feels the deepest wound in not instantly presenting himself for trial: but he knew not how Elberta stood affected – could she bear the shock of the trial?

She could bear all that virtue could demand – that affection could wish –

But – would her Children be allowed her? – Arnulph adores – but trembles in accepting her noble consent – trembles most for the Children, and wishes first to convey them safe from Offa's perjured cruelty – then present themselves –

El. 235 provides a slightly more detailed, cancelled summary:

She flies to welcome him. His sadness represses. He asks for Babes. She says they want food – He starts and will fetch – Wilfrid is expected. She asks his history? He was stopt by a messenger from Envoy, who told him he was implicated by Offa as a Traitor, he can scarce pronounce the word for indignation and inveighs against treachery, and animates for loyalty with fervour. His impulse was to present and vindicate himself, and brave his perjured accuser but . . . he must then either have proclaimed where were wife and Babes, or have abandoned them to no known rescuers: and Envoy tells him the danger would be great of generally delivering themselves up, such is the power of Offa.

He means but to put them under the protection of Envoy, and then, by delivering himself up, prove his innocence, and demand justice for them all.

Scene iii

ELBERTA, ARNULPH, WILFRID.

ELBERTA. O Wilfrid! give me food! –
O Wilfrid! have you brought my Children food?
WILFRID. Have you not Eat?
You look indeed most pallid –
ELBERTA. O for me –
But for these little ones –
If I am weak
What may their little frames endure?
O my good Wilfrid!
And comst thou empty Handed? Heav'n!
Heav'n have mercy on my poor Children!

Scene iv

Scene iii: El. 139 provides a brief summary:
Wilfrid enters sad and worn – She flies to him for food – he has none! terrified, she runs to see how her Children support themselves, beseeching Him to procure some at all events.
IV iii 1 s.d.–7 *ELBERTA ... frames endure*: from El. 140. The speakers' names are editorial insertions.
IV iii 8–10 *O my good ... poor Children*: from El. 141, which is marked A4 (i.e., Act 4).
IV iii 10 *poor Children*: El. 141 verso, also marked A4, contains the following cancelled lines:
ELBERTA. Where is the food? – O Wilfrid! give me food! –
WILFRID. In, in and hide thee! thy poor life is forfeit –.
Scene iv: El. 142 and verso provide a lengthy summary:
Wilfrid relates, he was met by a messenger from Envoy, who told him the flight of Elberta had occasioned her, through Offa's representation, to be pronounced guilty: and there was no doubt she would suffer if taken perpetual captivity, if not death.
Inexpressibly wretched, to submit to this or to the stain on his own Honour by absconding with her! A cruel cry for justice! – Says they will go forward, and brave oppression and falsehood.
But Envoy charges them all to fly to Scotland[1] till he can give them better[2] relief and counsel.
Then resolves he will leave her and Babes, and deliver up himself, to prove his innocence, and make a cry against her accuser's villainy.
Wilfrid says – how live, then? he had observed a person who hid himself[3] hard by – he suspects it to be Informo – though disguised – he durst not go on, seeing himself dodged – nor yet durst give the alarm to Elberta – Arnulph praises him –
Tenfold misery – durst not leave Wife and Babes unprotected if followed by that wretch – yet his honour hurt by every minutes delay.[4] He now must stay to see them safe –
Wilfrid owns he has not delivered Casket, and begs its use – high anger and refusal[5] –

ARNULPH, WILFRID.

WILFRID. See and forgive my zeal, my noble Master!
See here the . . .
ARNULPH. Quick take it back!
O tempt me not!
WILFRID. Thy wife, my lord, thy Children
ARNULPH. 'Tis true
The cause appears the cause of Heaven:
To nourish those who owe to me existence – Guard
The helpless innocence of infancy
Protect
I dare not go to look at them!
WILFRID. Now is the moment of auspicious time,
Now, when this booty –
ARNULPH. I gave my Honour to repose it safely –
Were it his own – twice nothing – he would joy
The means were his to save
But – a trust! – repos'd, too, by a man he hates –
Nor serves, but as a subject of the State
Officially, not friendlyly. Therefore more sacred
In that no social tie claims partial favour,

Wilfrid hardly presses his seizing it for support – He durst not – he says 'tis illegally Offa's – He hesitates – but resists – for the sake of justice to Ceolric for that of Honour to himself – No! he can clear it hereafter by saying he stayed to save wife and Babes – But a breach of trust is never rendered pellucid.

Wilfrid says they can no where be safe but with Malcolm.[6]

[1] 'fly to Scotland' is an insertion, replacing 'remain concealed'.

[2] 'give them better' is an insertion, replacing 'come himself to their'.

[3] 'he had observed . . . hid himself' is an insertion, replacing 'a wretch had robbed him'.

[4] 'Wilfrid presses seizing the booty' is deleted.

[5] 'Wilfrid owns . . . refusal' is an insertion, replacing 'and send Wilfrid on with Casket'.

[6] 'NB or instead of robbing – the proclamation' is deleted.

IV iv 1 s.d.–3 *ARNULPH . . . thy Children*: from El. 143. The speakers' names are editorial insertions.

IV iv 4–9 *'Tis true . . . look at them*: from El. 147. El. 148 offers an incomplete variant:
The cause, 'tis true, appears the cause of Heaven
So
The helpless innocence of infancy
Claimant of universal.

IV iv 10–12 *Now is . . . safely*: from El. 144, marked A4.

IV iv 15 *repos'd*: 'for him' is deleted.

IV iv 13–15 *Were it . . . he hates*: from El. 145, which continues:
Who sooner, therefore, may suspect design –
A man he execrates –
While he who does the deed, appears not in it –
But casts the semblance of black treachery
There where

But duty, just though hard, demands each sacrifice
Each strict observance – In
The Public Good, that cement of society
Pure Patriotism
Is deaf alike to enmity or friendship,
Impervious or to flattery, or resentment,
Or interest –
And all is apathy
Lest in one passion for the Public weal
One sublime passion the whole man absorbs
The Public good.
Nor Friend he seeks to serve, nor Foe to curb.

Scene v

ARNULPH, WILFRID, ACCA.

WILFRID. When the small portion it but –
ARNULPH. Ah Heaven! – well – take it hence – buy – purchase – sell –
Do what thou canst – save but their precious lives –
And only pray – for me, good Wilfrid – pray
Madness or Death relieve me! –
WILFRID. Many and many a fair fine day, I hope,
Will yet pay back this sadness with kind recompense –
But for their lives – once gone – who may recall them?
The Guide, unpaid before-hand, will not take them.
ARNULPH. Honour?
Avaunt, thou phantom! – Nature take me wholly! –
Yes! my sweet little ones! yes, my lov'd wife
Take – take the Casket – fly – this instant fly –

rushes out.

IV iv 25 *interest*: the following lines are given as an alternative:
Resentment personal, or private interest
Shut up each avenue to

IV iv 16–30 *Nor serves … to curb*: from El. 146, which begins by repeating 'a man he hates'.

Scene v: El. 149 provides a summary:
Coming hastily[1] she exclaims[2] at sight of Wilfrid, and reproaches him bitterly for not seeking the food the Children are pining for – Arnulph starts at the information, exclaims a wild defiance of false Honour, gives the Casket to Wilfrid, and hurries[3] him off – now to get food – then to convey all to Malcolm.
[1] 'back' is deleted.
[2] 'starts' is deleted.
[3] 'out of sight' is deleted.

IV v 1 s.d.–9 *ARNULPH … take them*: from El. 234. The speakers' names are editorial insertions.

IV v 10–13 s.d. *Honour … rushes out*: from El. 150.

IV v 13 s.d. *rushes out*: 'the Cave' is deleted.

Scene vi

ARNULPH. I will not tell her –
I will not sully her pure sense of right,
Lessening her hate of guilt – that guilt being mine!
Nor will I owe to pity that fair love
Which now with nobler Honour is my mede. –
O my Elberta!
Thy love for me would forfeit half its charm
Did I not know the high esteem which fosters it!

Scene vii

ARNULPH.

ARNULPH. 'Tis done – I have sent him for provisions –
And here is what remains of the rich prize
As yet by stealth untouch'd – By stealth? High Heaven! –
By Stealth? – of whom? – of Arnulph? death and torture!
Destruction – Infamy! – a Robber! – oh! –
A Robber? – hah! –
For you, my little ones, – for thee, my Wife,
Ye now may live – though never more sweet peace
Shall visit this torn Heart! O fiend like Honour,
How dost thou torture with fastidious gripe

Scene vi: El. 239 provides a brief, though slightly different, summary:

ARNULPH, ACCA.

Charges her to say nothing of the casket to his wife. Amazed she promises and goes in. The dialogue, from El. 234 verso, while not addressed to Acca, seems to discuss the same topic.

Scene vii: El. 236 provides a summary:
Agitated, he rejoices to save Children but starts from the means – cannot name Ceolric – calls himself a robber – and with loud exclamation utters bitter self reproach.

IV vii 1 *'Tis done*: El. 235 verso offers an alternative:
What I have done –
Necessity alone must justify.
Out! out!
Has Arnulph done what needs apology?
This is, indeed, some what that grates me home.

IV vii 1 s.d.–5 *ARNULPH ... Robber! – oh! –* : from El. 233, marked A4.

IV vii 6–17 *A Robber ... mercy*: from El. 207.

Thy erring votaries! away – away –
Hold me no more in thy preposterous fetters –
Give me, since suffering through my Heart's rich pores
Nature's severest pangs, give, give me respite –
Free me from sophystry of human weaving,
Imperious, adamantine, hard, inflexible,
And cast me wholly on soft Instinct's mercy.
Well! I have now got rid of this fine bubble
Of transient dazzle,
And I may live the better – and the wiser
That now I tread the Earth devoid of Shackles
Natural man
Uncumber'd with prescription's manacles
And the vain trammels of fantastic prejudice.

Scene viii

ARNULPH, ELBERTA.

ELBERTA. What noise is that? Wake not my little ones –
Again I have hush'd them
Sweet little Lambs! they can but wake to weep,
And weep again till Slumber brings them peace.

IV vii 19 *Of*: an insertion, replacing 'This'.
IV vii 22 *Natural*: before 'Natural', 'A' and an illegible word are deleted.
IV vii 18–24 *Well! ... prejudice*: from El. 232.
Scene viii: El. 231 provides a brief summary:

Elberta comes forth to beg him not to[1] awaken his Children – He tries to compose himself – she asks how long they may live without nourishment? deeply affected he sends her in –

In El. 230, Elberta asks how long the children can live without food but addresses Acca rather than Arnulph:

ELBERTA. Tell me, Acca, truly,
How long can Nature struggle without nourishment?
How long can such weak objects bear the gripe
Of knawing Famine, ere to rest it sinks them! –
To rest! – for Oh! sweet rest, soft downy rest
Awaits their parting souls – Who knows how soon?
O who shall count the labouring – fleeting minutes
So slowly sad while passing, though so brief,
Once o'er, to recollection and regret?

The speaker's name is an editorial insertion.

[1] 'visit' is deleted.

IV viii 1 s.d.–4 *ARNULPH ... brings them peace*: from El. 229.

ARNULPH. Give me in silence credit for my love –
O if thou knewest to what extent its fondness
For thee and thine could urge me! –

Scene ix

ARNULPH *solo*.

ARNULPH. Factitious Honour, hence! hence, captious Reason!
Ye schools, ye sophists hence! Declaimers subtle!
Spin fine your wire-drawn texts, perplex, confuse
The face of things, invert fair nature's order,
Teach feeling, apathy, defy sensation,
Scare from her conscious rights all native simpathy –
Refine, reduce, annihilate the senses –
And then – what have ye done for frail humanity?
What are your tenets to a Father's Breast?
A Husband's? To each charm of social life?
To one sweet smile of springing Infancy?
To one endearing look of natural love? –
O argument of soul! whose genial laws
Level, yet elevate all human kind!
Whose silent powers entwine the captive Heart
Resistless, urging, or, persuasive, melting,
Thine be my energies! to thee devoted
My faculties, my life, – eloquent Instinct!

IV ix 2 *schools*: schoolmen or scholastic philosophers; an anachronism.
IV ix 2 *sophists*: those who use fallacious arguments; specious reasoners.

IV viii 5–7 *Give me in silence* ... *could urge me*: from El. 228. The speaker's name is an editorial insertion.
Scene ix: El. 227 provides a brief summary:
Parental fondness reconciles him to what he has done – he disdains vain cavil – and deprecates factitious Honour for Nature's nobler Instinct.
IV ix 10 *charm*: an insertion, replacing 'joy'.
IV ix 1 s.d.–11 *ARNULPH solo* ... *springing Infancy*: from El. 237, marked A4.
IV ix 12–18 *To one endearing* ... *eloquent Instinct*: from El. 237 verso.

Scene x

INFORMO.

(*Changes to Cave*).

INFORMO. I do not zest employment against Children
I thought it nothing when he gave the office –
But when I see them . . . I could find i'my Heart
To foster sooner than despatch . . .
Would Arnulph could reward me for betraying him –
Yet – now to fail . . . I lose my mede for all –
I will not look these Children in the face
Again . . .
Perhaps, at last, he may not murder them –
If he to me should give the dire commission
I'll get some other to perform it . . .
I'll find some starving wretch – such as myself
Ere Offa gave me pay . . . yet then I slept! –
Yes, on the frozen Earth – or burning sand –
Unfed – unhoused – uncherish'd – I could sleep
When wearied Nature call'd for soft repose –
And now . . . to close my Eyes becomes a curse
I shout – see horrid phantoms –

Scene xi

ELBERTA *sola*.

ELBERTA. To common Eyes they look no more so beauteous –

Scene x: El. 225 provides a summary:
He glides in, delighted at length to have discerned the abode – yet perplext how to get Children, and vext at the office. Prefers employ against men – less innocent and unoffending – Somewhat laments his own loss of what resembled them – innoxiousness – Again hopes when rich to expiate all. Must watch his moment of security for entering the Cave which will be night –
The dialogue is from El. 224.
IV xi 1 s.d. *Scene xi*: beginning with this scene, all scene numbers are editorial insertions.
El. 223 provides a summary:
She sees, trembling, her Children fading – she sighs, but submits to[1] what must bring them to be angels. A noise alarms – should it be pursuit.
The dialogue is from El. 222, marked A4.
[1] 'thus' is deleted.

To mine they are all the dearer for the change.
Yet, as I look at them, – and see them fading –
All paler turning – losing from their Cheeks
Those rosy tints that, fresher than the morn
Salubrious glow'd, – drooping their little Forms
With secret pining at their alter'd life,
Which, all unknowing why, impressive proves
Ev'n to their thoughtless Breasts, of sadd'ning simpathy,
Dear, innocent, sweet sharers of my woes!
O yet for you be happier Days in store!

Scene xii

ELBERTA, WILFRID.

Scene xiii

ENVOY *and a party of soldiers*.

ENVOY. She's sure in this vicinity –

Scene xii: El. 221 provides a summary:

He comes with food – but says a party of soldiers are in pursuit he fears, and bids her hide – she charges him to stay by her little ones within, and guard them – and she will watch without – She makes him fly in with food – then shivering with terror, she enters the mouth of the Cave!

Should they rudely enter, it may affright[1] them to corrosive terrors –

There is no dialogue for Scene xii.

[1] 'their little' is deleted.

Scene xiii: El. 220 provides a summary:

Envoy says this must be near the spot – a soldier marks the mouth of the Cave – They are advancing – Elberta rushes forth – and says her life defends all[1] entrance. They would brave – she says she is a mother, and defies them. They listen with respect – And Envoy challenges her – she owns herself, and desires to see him in private.

The soldiers retreat and she supplicates he will find means to let her remain – He says he knows her marriage, and she had better remain, as Arnulph is in danger – she protests they will all live or die together – he consents to let her pass for another, and come himself to remove her and Babes by dead of night.

He says he found she had escaped from Ceolric, and his command was to find her, under high penalty, having been surety for her faith and loyalty.

Arnulph is strolling she knows not where.

[1] an insertion, replacing 'their'.

IV xiii 1 s.d.–4 *ENVOY and ... evil purpose*: from El. 215. The speaker's name is an editorial insertion.

We trac'd her from the Cottage on the Moor,
And think her to be spirited away
For evil purpose.

ELBERTA. Stay! Here if ye come –
Behold your passage! (*opening her Bosom*)
Here, if ye force your way
This Breast must first be mangled, every Limb
Be dislocate, and my life's blood in torrents
Rush from each vein to overwhelm my murderers.
No! here ye enter not!

ENVOY. And who resists us?
A woman?

ELBERTA. Yes! that woman is a mother.
A mother guardian of her Children.
Know ye what may be stronger?
Is't not to such ye owe your proud existence?
And think ye such to brave?
No!
There is a voice innate to every Heart
That robs your pride of force, your power of strength
Though still and silent in the inward Soul
Man's haughty policy refuse it utterance
That more than supplicates, commands respect
For the fond mother who protects her Children.

ENVOY. Why thou art valiant –

ELBERTA. Were I a Lion rushing from this den
Could I be more? Think ye the cries of famine
Can sharper nerve the frame – or pierce the Heart
With fangs more potent than the tender voice
Of Childhood crying for support and shelter?

ENVOY. Well, can ye give us tydings, noble Dame
Of our fair captive Princess? Five long years
Elberta of East Anglia
Missing to all, we deemed her midst the slain
Of late we hear –

IV xiii 4 *Stay*: an insertion, replacing 'Hold'. 'Here if' is deleted and recopied on the next line.
IV xiii 9 *Rush*: 'forth to' is deleted.
IV xiii 11 *that woman ... mother*: 'A woman and a mother' is given as an alternative.
IV xiii 13 *Know ye*: before 'Know ye', one or two illegible words are deleted.
IV xiii 4–22 *Stay ... protects her Children*: from El. 218. The speakers' names are editorial insertions.
IV xiii 27 *tender*: an insertion, replacing 'babes soft'.
IV xiii 28 *support*: an insertion, replacing 'relief'.
IV xiii 23–28 *Why thou art ... support and shelter*: from El. 219. The speakers' names are editorial insertions.
IV xiii 29–49 *Well, can ye ... gripe the Heart*: from El. 216. The speakers' names are editorial insertions.

ELBERTA. What would ye with her?
ENVOY. Sole offspring of our loved and murdered King
Whose memory still must live, while goodness, honour,
Benignity and mercy pass for virtues
August and Kingly . . .
I'd give my very life to ransom her –
SOLDIER. And I –
ENVOY. Yes, the whole Kingdom sighs to do her right
In reverence for
ELBERTA. Is't even so? Know then, ye generous men
Elberta of East Anglia – stands before ye!
Ye start? Ye know me not?
Alas! the changes made by time are naught
To those of sorrow!
The deprivation year on year
What is it to the harrow of Affliction
Where Terror joins with Grief to gripe the Heart?
ENVOY. Noble Creature say'st Thou, so
If such thy charge – for all that man can give
Or all that man can seize in victory's Car
I would not injure Thee! –
If here thou durst to dwell –
With our own patrole
We'll guard thy habitation –
ELBERTA. Good gentle sir –
Let me then speak with you alone
I pray you find some means to leave me here
I wait a friend – one whom, perchance, in missing me
May deem me lost . . .
ENVOY. Lady, if 'tis thy husband – know –
ELBERTA. Think you I'll save myself by losing him?
No! we will share alike –
All Affluence were sorrow –
I'll work – I'll labour –
My little ones I'll breed to hardest toil –
If hard the toil that keeps them with a Father –
O such a Father! –

IV xiii 50–56 *Noble Creature* ... *habitation*: from El. 217. The speaker's name is an editorial insertion.
IV xiii 57–67 *Good gentle* ... *such a Father*: from El. 214. The speakers' names are editorial insertions.

Scene xiv

WILFRID.

WILFRID. O divine Nature!
How wondrous are thy workings!
'Tis not a minute since, feeble, subdu'd
Feverish her habit, tremulous her frame,
She sunk – desponding – helpless –
Scarce keeping strength to bear shaking Limbs
Behold the change!
Danger now pressing for her little ones
Behold – not strength restor'd – but vigour fresh,
And force, and energy – of sudden growth,
Till now unknown! And intrepidity
Mental, of equal novelty and force,
Turning aside
The gentle stream that timid flowed
Through every vein – for the rough, tumbling torrent
Of more than manly force –

Scene xv

ELBERTA. Yet – didst then –

Scene xiv: El. 213 provides a brief summary: 'He leaves cave as Elberta enters it, marvelling at the courage given by maternal feelings'. El. 188 provides an even terser summary: 'Wilfrid. amaze and alarm'. The dialogue is from El. 212. The speaker's name is an editorial insertion.

IV xiv 5 *sunk*: 'helpless' is deleted.
IV xiv 6 *keeping*: an insertion, replacing 'seeming' and an illegible word.
IV xiv 6 *bear*: 'her' is deleted.
IV xiv 8 *now*: an illegible word is deleted.

Scene xv: El 198 provides a summary:

She comes to seek – he turns from her in anguish silent and gloomy – says Well – have you fed your Children? says 'tis with their Fathers Curses – his Infamy – Quits her amazed –

The dialogue is from El. 197 and verso. The latter is marked A4. The speakers' names are editorial insertions. El. 196 and verso provide slightly different but related dialogue. El. 196 is marked A4. The speakers' names are editorial insertions:

ARNULPH. Come – feed – here's nourishment –
ELBERTA. Feed too –
ARNULPH. I cannot – 'tis so long I've fasted –

ARNULPH. Why dost thou probe me thus with artful question?
ELBERTA. My Love?
ARNULPH. Away! away! – ask me no more
Be my last portion ignominy –
Be dire remorse the pillow of my exit
Hopeless my latest sigh
If ere again I answer these tormentings.
ELBERTA. What have I said? O my poor Heart will burst.
Why this disorder? O my Love! my Husband,
Stare not thus wildly? – If 'tis I have hurt thee –
Pardon my ignorance! – O look not thus! –
Lord of my Heart! behold me at thy feet –
ARNULPH. Why? – what? wouldst make me mad?
Your lives –
Made of your Father's infamy – no matter; –
Ye live! – my guilt
Will not descend to you – my woe – my poverty –
And what of those? Lighter than the pure air
Playing in Zephyrs round the Mountain tops
Compar'd with the dense atmosphere of guilt –
The thick, corrupted streams of fell remorse,
Exhaling worse than pestilential fumes
From the polluted conscience – Nay – go in –
Why stayst thou in the cold? – go in – go in!
ELBERTA. Alas! I must obey! – but O! – be calmer –

ELBERTA. Prithee be cooler – why art thou thus impatient?
ARNULPH. My[1] best love?
ELBERTA. Ay – werefore thus disturbed? What should alarm thee?
ARNULPH. And ask me not question – pray mark that!
Have they both Eat?
Then have they eaten up their Father's curses!
Ha![2] – nay, heed me not – I did but jest –
ELBERTA. O jest not in such sort!
ARNULPH. My love, I will not. Calm thy ruffled spirits –
Be all s‹eren›e – see how but it unhinges me
To see weep –
No! weep, weep, weep! – Celestial Tears! – they flow
From snowy sources of soft guiltless quiet!
I cannot weep such! – should I see thee die
And both thy little ones hang o'er thy tomb,
It would not move me to one Tear like thine!

[1] before 'My', a word ending in 'itle? – ' has been torn away.
[2] before 'Ha', a word ending in 'n!' has been torn away.

Scene xvi

ARNULPH, WILFRID.

Changes to a Grove.

ARNULPH. Tell him to send me . . .
Say if he will trust . . .
Trust? – ha! – trust, did I say? –
The hot blood mounts my Cheek with burning shame
Accursed moment! am I then so lost
I dare not beg – unblushing – to be trusted?
Had I announced their wants – I might have died –
And what of that? They would have been alive
And I – a happier man, because an innocent
Had died unsullied – therefore died content!
If I had died – 'twere o'er – how soon forgot
Ashes to ashes mix'd, and dust to dust –
But here – O Infamy survives –
Ay – but they too had died! –
Well! Heaven had then received them but the sooner.

IV xvi 12 *Ashes to ashes . . . dust to dust*: echoing the Committal in the Order of the Burial of the Dead in the Book of Common Prayer of the Church of England: 'earth to earth, ashes to ashes, dust to dust'. The sources are biblical: 'till thou return unto the ground; for out of it wast thou taken: for dust thou art and unto dust shalt thou return' (Gen. 3: 19) and 'then shall the dust return to the earth as it was' (Eccles. 12: 7). The exact phraseology is, however, peculiar to the Book of Common Prayer and, hence, anachronistic here. See Stella Brook, *The Language of the Book of Common Prayer* (London, 1965), p. 200.

Scene xvi: El. 178 provides a summary:

He is sending Wilfrid to Ceolric, to relate the History, and say he will try to replace all in happier days, if from his dire necessity he will trust.

Starts at word trust – and will not send at all –

Wishes he had announced his Babes distress, and died for them.

Sees the plain road at last the only one that has chance of success.

Wilfrid tries to console[1] – he does not hear – cannot listen to him – but goes on sadly moralising, now for Honour, now for Instinct.[2]

Refuses to even look again at Casket –

How indignant must be Ceolric, who will be accountable – and can only clear himself by impeaching his friend – What a friend! – O disgrace!

Shame! Infamy! –

Sends on Wilfrid –

Wilfrid has heard his name is struck from the list of officers – and posted for a deserter.

[1] 'A messenger comes – that Ceolric' is deleted.

[2] 'Says he must respect' is deleted.

IV xvi 1 s.d.–6 *ARNULPH, WILFRID. . . to be trusted*: from El. 211.

IV xvi 7–10 *Had I announced . . . died content*: from El. 208.

IV xvi 11–15 *If I had died . . . the sooner*: from El. 203.

Would I had rather suffered them to starve –
O no! o no! – a thousand Deaths were nothing –
Exposure – Shame – Dishonour – Ignominy –
What are ye all to one such innocent life?
Ye tender pledges of my Heart's dear love!
Ye soft Entwiners round my soul's affections,
Live – and be blest! Live to console – restore
The hapless Partner of my life's sad fate –
The Pattern of all worth –
How clear – but o, alas! how late I see
The line of duty still is strait –
Had I sustain'd my port –
Stood firm against temptation and distress . . .
O what of bliss were mine by this soft amnesty!
It may not be! I have lost the purpose of life,
The spur of action, the fair pride of Fame,
All that supports the inward man, that dictates
Repelling courage to assailing scorn,
Contempt of poverty, firmness in woe –
O Honour! Thou art then at last no phantom
Unquestionable tie of man to man,
Unbidden monitor of human claims
As conscience of divine! thou noblest spur
To generous efforts, casting still behind
All lurking littleness, each narrow feeling
And all the base prerogative of self –
O not alone the bond of man to man,
Man to himself thou bind'st, by stronger chains
Than law can forge, authority, or arms,
Entangling ev'n volition! – chains impervious
To vulgar Eyes, but to the Just transparent!
Alas – the trammels of the World's prescriptions
Vainly we strive to scatter and disband!
Honour! –
O lost to me! – Vain now were fortune's smiles –

IV xvi 27 *port*: behaviour, conduct.

IV xvi 16–24 *Would I had . . . of all worth*: from El. 204, marked A4.
IV xvi 28 *distress*: 'Now might this amnesty' is deleted.
IV xvi 25–29 *How clear . . . soft amnesty*: from El. 209.
IV xvi 30–34 *It may not be . . . firmness in woe*: from El. 206, marked A4.
IV xvi 39 *behind*: 'Each narrow feeling' is deleted.
IV xvi 40 *All*: an insertion, replacing 'Each'.
IV xvi 35–41 *O Honour . . . prerogative of self*: from El. 210, marked A4.
IV xvi 42–46 *O not alone . . . Just transparent*: from El. 210 verso.
IV xvi 47 *Alas*: before 'Alas', 'Come to the Children come, if' is deleted.
IV xvi 47–55 *Alas . . . internal war*: from El. 205, marked A4.

Prosperity unacceptable –
Then still, calamity, retain thy bitterness!
Mock me not now with one vain pitying turn –
What can the world of peace – ease – joy – import
To him whose Mind wages internal war?

Scene xvii

OFFA *solo*.

Scene xviii

CEOLRIC, OFFA.

Changes to the Tower.

CEOLRIC. He merits to be tortur'd on the Rack.

IV xviii 1 *the Rack*: fifteenth-century instrument of torture consisting of a frame with a roller at one or both ends to which the victim's legs and hands were tied. By means of pulleys and levers, the ropes were tightened until the victim's joints were dislocated. Its mention here is an anachronism.

Scene xvii: there is no dialogue for Scene xvii, but El. 238 begins with the following cancelled summary:

Informo says he discovers they have the Casket. Offa says this will do, for he can bring them to ruin through their friend – he will demand it of Ceolric – and force him to impeach Arnulph – who then may pay the forfeiture of the felony, and so rid him of such a rival at once and openly.

Informo objects [an illegible word].

The rest of the summary is uncancelled:

No casket delivered! – is Ceolric a felon? has he discovered the Treasure was of Ethelbert and purloined it for Elberta? He will not undeceive Informo, who is so mercenary, he might refuse to act, and who merits the disappointment – but he will seize Ceolric, and recover his property, or be avenged by his life with stamped dishonour.

Out witted by a thing that passed for a dupe to every Knave he met with! – O intolerable! –

Suspects tis taken by Ceolric to support Elberta in her flight – and her progeny –

Out! out! He will seize him –

El. 6 provides an alternative cancelled summary:

Informo brings word the Casket has never been delivered at the Fortress. Enraged, he rushes to claim it of Ceolric. Fears he has opened, and detains it as Elberta's property – resolves to protest he meant to secure it as such for her use upon leaving home.

Resolves to watch for seizing Children himself – he is unknown to Acca, with whom they are left, while Elberta and Arnulph wander, and Wilfrid seeks food, as Informo says.

Scene xviii: El. 226 provides a summary:

Offa demands Casket.

Ceolric answers 'tis delivered to Fortress.

He denies it. Ceolric enrages. He proves it is not there. Ceolric becomes confounded.

OFFA. Who?
CEOLRIC. No matter.
Had I not seen those helpless little ones –
OFFA. What sayst thou? –
CEOLRIC. O I could see him toiling in the mines,
Consuming midst
Writhing with agony in
Thus to betray . . .
OFFA. I see he's dup'd – for on his brow is writ
Something of deep concern amidst his anger
That argues earnest disappointment;
Such as befits a novice in deceit.

Scene xix

ARNULPH, COTTAGER, ROWENA.

He accuses him of guilt. He storms in his own defence. Offa says who then, what, or how? Clear yourself. He grows sullen. He says finally, he has taken measures coercive; he must produce the treasure, or be arrested. He is arrested.

Unhappy, guilty Arnulph! he exclaims, I cannot forgive this treachery! – thou wert a friend, and trusted – But thou art a Husband – a Father – distressed and persecuted – and I cannot betray thee!

Offa, accused it was treasure of Ethelbert, says yes, it was to guard it for Elberta.

El. 178 verso offers a similar cancelled summary:

Scene at Tower

Offa claims casket of Ceolric saying it is not at Fortress. Ceolric enraged – will not give up Arnulph.

Offa says must deliver it, or be arrested –

He is arrested.

IV xviii 1 s.d.–8 *CEOLRIC, OFFA . . . betray*: from El. 241. The speakers' names are editorial insertions.

IV xviii 9–12 *I see he's . . . in deceit*: from El. 262.

Scene xix: there is no dialogue for Scene xix. El. 202 provides a summary:

Arnulph, perceiving Cottager and his wife, retreats.[1] She says she must take breath, he answers they may be too late, . . . speak of the grand Trial,[2] of pity great folks should be thus rapacious – never satisfied – however splendidly superior – name dishonour[3] and a General – They are going –

Arnulph ventures forth – and asks what Trial[4] they speak of! The Generals. What General? Ceolric. Great Powers! What is his crime? He purloined rich treasure of Jewels and Money from Offa –

Thunderstruck, he says They lie! –

Affrighted, they retreat –

[1] 'They' is deleted.
[2] an insertion, replacing 'procession'.
[3] an insertion, replacing 'Treason'.
[4] an insertion, replacing 'execution'.

Scene xx

ARNULPH.

ARNULPH. What him? the noblest Friend that man e'er own'd –
The soul of goodness – Hah! for me – for mine –
Forfeit his precious Fame? For deeds of mercy
Be paid with Ignominy? For pity's softest kindness
Sustain the stigma of foul treachery?
Of Treason? – of dishonour? –
What may he say? – All ways he is lost.
'Twas me? – how came I by't? –
No, he must seem himself the violater
Or me betraying, be yet more betray'd.
And is it thus I pay the generous daring
His friendship urg'd? the nobly trusting courage,
Tribute to virtuous faith in spotless Honour.
O I am sick within! – This cannot prosper! –
No! – I must speed to him what I have done!
Ah! punishment severe the degradation
Of such avowal! – words are wanting to convey it
Without disgrace! New Language must be coin'd.

Scene xx: El. 201 provides a summary:

In deepest anguish at news of seizure of Ceolric,[1] resolves to avow instantly, and save him from own suspicion – yet his wife[2] – his Children – yet – he must go and know the whole, whatever be the risk –

The dialogue is from El. 199–200. El. 199 is marked A4.

[1] 'resolves debate what to do?' is deleted.

[2] 'His wife [an illegible word] – [an illegible word] he has not [three or four illegible words] how' is deleted.

IV xx 3 *Fame*: an insertion, replacing 'life'.
IV xx 4 *Ignominy*: an insertion, replacing 'death'.
IV xx 1 s.d.–6 *ARNULPH ... of dishonour*: from El. 199.
IV xx 13 *Honour*: 'held undoubted as invulnerable' is added in pencil.
IV xx 7–18 *What may he say ... be coin'd*: from El. 200.

ACT V

Scene i

ARNULPH.

Cave.

ARNULPH. When I behold that cave – the gloomy refuge
Yet sole receptacle from Death – from Treachery –
Where my sweet Children – artless of all wrong
Yet to all ill expos'd –
Yes – now suspense is o'er –
But dire – not soft the calm it leaves!
Well! – be it as it may for happiness
My Nerves and fortitude once more are brac'd.
My Tale is told – I die.
One little doubt alone is worth a balance.
Shall I or seek – or shun a last adieu?
My wife – my Babes –

Act V: El. 264 provides a summary entitled 'Scene Grande'. The first five lines describe events occurring earlier; the remaining events take place in Act V:
Elberta rushing from Cave to guard Babes.
Arnulph meeting unprepared his wife at Offa's.
Offa surprising Elberta and Arnulph.
Arnulph's entering disguised to Ceolric at Tower.
Elberta and Wilfrid forcing Cottage doors upon Arnulph.
Arnulph rushing forth to arrest execution of Ceolric.
Elberta avowing herself to the troop –
Elberta entering at Arnulph's dying scene.
Elberta rushing forth to demand her Children.
Wilfrid seizing the Dagger of Informo.
Arnulph discovering himself to Envoy –

Scene i: El. 191, marked 5 (i.e., Act V), provides a summary:
He resolves on Death – to save Ceolric.
He looks tenderly at the Cave – he hesitates whether to take leave or not – at length resolves to bless himself with one last embrace – but not to let them know tis final but leave his sad tale with Envoy for their vindication and refuge.

V i 1 s.d.–4 *ARNULPH ... ill expos'd –* : from El. 189, marked A5. The speaker's name is an editorial insertion.

V i 11 *adieu*: two lines are deleted:
My wife [an illegible word] I must not think after
My Babes – let me take them from my thoughts.

V i 5–12 *Yes – now suspense ... my Babes –* : from El. 190, marked A5.

Scene ii

ARNULPH, ELBERTA, WILFRID.

ARNULPH. We sure shall meet again –
ELBERTA. Again? – What means my love?
ARNULPH. I think so!
ELBERTA. You but think it?
ARNULPH. Nay – I trust!
ELBERTA. You trust?
ARNULPH. Yes – in high Heaven! –
Exit.

ELBERTA. Come! to thy Children come – if not to me!
Live! for thy Children Live – if me thou hatest!

Scene iii

ARNULPH, WILFRID.

Scene iv

ARNULPH *solo*.

ARNULPH. O for a moment now to man myself

Scene ii: El. 187, marked 5, provides a summary:
He tears himself away from her – unexplained – but most tenderly – blessing her and her Children – charging her not to follow – and prepare for beginning the journey to Scotland at night – Then turning round – Who, he says, can carry little ones? – ah! – forces her in –
The dialogue, in which their parting is not as tender as the summary suggests, is from El. 185–86. El. 185 is marked 4.
V ii 3 *Heaven*: 'Mean you we meet – or that you trust in Heaven?' is deleted.
V ii 1 s.d.–3 s.d. *ARNULPH, ELBERTA ... Exit*: from El. 185.
V ii 4–5 *Come ... thou hatest*: from El. 186.
Scene iii: there is no dialogue for Scene iii. El. 183 provides a summary:
Arnulph charges him, at all hazard, to precede him to Envoy, with whom he must speak – and to conduct his wife and Children by dead of night toward Scotland.
Scene iv: El. 182, marked 5, provides a brief summary:
He arms himself with fortitude for death after unconscious leave-taking of his Wife and Children – says he must to Envoy on secret purpose previous –
The dialogue is from El. 181, marked A5, and El. 192, marked for either A5 or 2.4 (Act II, Scene iv).

From the sharp scenes of exquisite distress
So tearing to my soul! – and yet so dear! –
How little think they that these wretched arms
Have giv'n their last embrace! Yet how they wept!
Unknowing half their woe! And I the wretcheder
Knowing the fatal all, to see their ignorance.
Yet – had they known me as I am – just going –
A parting Husband – and a dying Father –
How had I borne the tender last caress
That riv'd my secret Soul, thou giv'n
Let me not dwell on them! – Come, higher
Come, energy! to noblest Friendship due!
Come, Firmness! Honour! Gratitude! and Truth!
Since these demand the sacrifice of Life.
O Death! Thou solemn refuge of Despair!
How would this wearied frame solicit thee,
Call for thy rest to still its grievous sufferings,
Quiet this throbbing Pulse, these feverish pains,
Were not the inward agony of soul
Ev'n yet more dire! an agony that cries
Loud to the Breast maternal Die not yet!
Support thy labours, prop thy aching Limbs,
Think what thou leav'st – and fear for what thou sav'st!

Scene v

ELBERTA *sola.*

V iv 3 *So*: an insertion, replacing 'Most'.
V iv 5 *giv'n*: a word beginning 'unc' is torn away.
V iv 12 *higher*: a word is torn away.
V iv 1 s.d.–15 *ARNULPH . . . sacrifice of Life*: from El. 181.
V iv 16 *Despair*: an insertion, replacing an illegible word.
V iv 16–24 *O Death . . . thou sav'st*: from El. 192.
Scene v: there is no dialogue for Scene v. El. 177 provides a brief summary:

Comes to Arnulph – he is gone[1] – alarmed and dismayed, resolves to endeavour to trace his footsteps, in fear and horror of some desperate resolution yet distantly, and unperceived – till she can meet with Wilfrid to follow him –

[1] 'Comes to Arnulph – he is gone' is an insertion, replacing 'Comes forth, and finds Arnulph gone'.

Scene vi

INFORMO.

Scene vii

INFORMO, ACCA.

INFORMO. Come hither, my good Woman –
Thou hast two doughty Children in thy charge –
Bring them me –
'Tis meet they change the air –
Seest thou not they are sickly?
ACCA. O alas! –
INFORMO. Well – bring them here.
ACCA. For why?
INFORMO. Bring them – and instantly.
ACCA. Me? – bring them? Never! –
How durst thou ask? Who gave thee such command?
INFORMO. Look at this little Instrument –
ACCA. O Heaven –
INFORMO. How dost thou like its mien?
ACCA. Ah! mercy! –

Scene vi: there is no dialogue for Scene vi. El. 176 provides a summary:
Comes forth, rejoicing to have seen Arnulph, Elberta, and Wilfrid severally depart – makes to seize Children without delay. He has a place prepared, secret, yet not distant, for their immediate reception, whence he[1] may wait[2] further directions with them – agitated and uncertain what they may be – There, he says, they may lie, while he claims of Offa his treasure, lest he should be disappointed – not trusting him except while he has power over him, and intends to possess before he removes the Children further, or obeys any commands.
[1] an insertion, replacing 'Informo'.
[2] 'his' is deleted.

V vi 1 s.d. *INFORMO*: an insertion, replacing 'OFFA'.

Scene vii: El. 175 provides a summary:
Just entering the Cave, as Acca comes out, calling upon Elberta to appease her little ones – He demands the Children of her instantly. She refuses – he shews her a dagger, and swears it shall be plunged in her heart if she utters a scream. He follows her into the Cave.
The dialogue is from El. 174. The speakers' names are editorial insertions.

V vii 1 s.d. *INFORMO*: an insertion, replacing 'OFFA'.

V vii 2 *doughty*: an insertion, replacing 'little'.

V vii 10 *mercy*: two lines are deleted:
INFORMO. Bring me the Babes!
ACCA. O do not – do not – .

INFORMO. What think you should it enter in your Breasts.
ACCA. O do not! – do not fill me thus with terror
What have I done? –
INFORMO. Bring me the Children!
ACCA. O no! o no! –
INFORMO. This instant – or receive
Deep in thy Heart this weapon! –
ACCA. O pity! pity! –
INFORMO. Bring them, I tell thee, woman! –

Scene viii

ENVOY, ARNULPH.

Changes to a Castle.

ARNULPH. Thou see'st before thee one – who . . . erst! – was deem'd
A Knight of spotless Name, and well tried valour:
A Knight of fame untainted, and of valour seeming
A man
To Honour unindebted; to reproach
A Stranger;
Who in the front of Day could meet his foe
Fearless of danger fight, of slander, pardon,
Such, – erst! – was . . . Arnulph!
ENVOY. Arnulph?
ARNULPH. He. Though now
A fugitive from man, and from society

Scene viii: El. 173 provides a lengthy summary:

He says – discovering himself, 'tis Arnulph, the unhappy Arnulph stands before him. He means to challenge Offa for his base calumny, and supplicates him,[1] should he fall to protect his hapless wife and Children – Envoy promises all and with honourable warmth of feeling – and will convey the lady and Children to Malcolm, till the Conqueror can hear the whole history – and receive them, with the rest of Atheling's friends who have surrendered, with[2] his same soulless self – (fitting for women and Children says Ceolric)

Arnulph calls upon him to second his honourable vindication, not mar it – and retires – Since he has the command at Ceolric's doom.

The dialogue is from El. 152, marked 5. The speakers' names are editorial insertions.

[1] 'as he is a man, and a soldier to forbear punctilio as a Governour' is deleted.

[2] before 'with', there is an opening parenthesis.

V viii 10 *society*: 'From man a fugitive; from fair society' is given as an alternative.

An outlaw.
Yet is he still the same – unchanged, unaltered,
No vilifying passion warps his reason,
No arts of sophystry debase his reason
No vilifying passions snare his Heart
Unwrapt his reason by degrading passion
Untouch'd his Heart by tenets of corruption.

Scene ix

ELBERTA, ACCA.

Changes back to Cave.

ELBERTA. Acca! why standst thou here? Thus all alone? –
Sleep my beloveds? – Thou lookest all aghast –
Are they not well?
ACCA. How ever may I tell these dreadful tydings?

V viii 15 *snare*: 'warp' is given as an alternative.
Scene ix: El. 172, 171, 169, and 168 provide brief summaries:

ELBERTA *sola.*

Enters seeking Husband, sick of Life.

ELBERTA, ACCA

She stands at head of Cave – asks after her Children – no answer – she rushes into the Cave.

ACCA

Bewails how to tell dire tydings of Children's seizure.

ELBERTA, ACCA.

Rushes wildly from Cave to demand her Children – runs off Acca vainly comforting and calming, pursuing her.
The dialogue is from El. 170.

Scene x

WILFRID, ACCA, INFORMO.

Changes to Grove.

WILFRID (*producing a pistol – craftily takes from Informo*).
Now then, villain! –
Most desperate, hardened villain! now deliver
Thy charge – where are the destin'd Innocents

V x 1 s.d. *pistol*: an anachronism by approximately five hundred years.

Scene x: El. 166, 165, and 164 provide a summary:

WILFRID.

Says sentence passed, and all preparing for execution of Ceolric. Dreads the effect upon Arnulph –

WILFRID. ACCA.

She asks if he has seen Elberta? No – she has fled, and seeing a crowd, plunged headlong into it, demanding her Children, and breaking through all ranks from velocity of frantic sorrow. He supposes she has joined the multitude that gather to see the dread execution – will hasten to protect her – asks why she seeks Children? Acca relates their seizure. She shrieks – 'Ah! there's the murderer! – let me hide for pity!' and runs aside –

WILFRID, ACCA (*apart*). INFORMO.

Informo is hastily crossing, saying he must instantly see Offa – when Wilfrid joins him – and pathetically demands Children. He denies all knowledge of them[1] – Wilfrid instantly offers him all the treasures of the Casket for the Babes' preservation – he exclaims[2] How – Is not the Casket at the Fortress? did Offa delude me thus infamously? Wilfrid adds, with energy, he cannot be so well paid any other way, as the Children must be immediately[3] pursued, and he cannot get his reward, and be off in time to save life and recompence. Informo enraged at Offa, resolves to take it, and betray him – asks if 'tis ready – Wilfrid says yes, and produces it – he looks at it – and finds it irresistible – he accepts it, provided he will go to the place where the Children are, and there remain till he can abscond – He believes he shall be better paid than by the rapacious and parsimonious Offa – Wilfrid pleads impossibility with two Infants to be speedy, even if willing. Informo says 'tis true – Wilfrid gives it – Informo, in putting it up, discovers the hilt of a dagger: Wilfrid, pretending to aid him, snatches it – he then bids him re-deliver the Casket, and shew him that moment to the Children, and accompany him for Confession[4] on pain of instant death – terrified he obeys – and Wilfrid calls on Acca to attend, and take Children, while he guards Informo.

The dialogue is from El. 163.

[1] 'but says Wilfrid has the Casket' is deleted.
[2] an insertion, replacing 'listens'.
[3] an insertion, replacing 'ultimately'.
[4] 'to the place of Ex' is deleted.

V x 1 *Now then*: an illegible word is deleted.

The victims of thy avarice and barbarity?
This instant yield them – or the breath of life
Which now protects thee from the Hell thou meritst
Cut short, shall

Scene xi

Scene changes to Parade before the Tower.

Scene xii

Scene xi: there is no dialogue for Scene xi. El. 162, marked 5, provides a summary:

Grand procession – drums – dead March –

Ceolric is on the point of Execution –

Arnulph rushes in, and declares himself culpable of the robbery of the Casket.

The soldiers would seize him – he resists – They are intimidated by his desperation – he attacks Offa, calls him villain, and insists on reparation for his honour and his wife's, that he may leave the world untainted by so foul a murder – He pathetically recommends to Envoy and to Ceolric his wife and Children.

Envoy would interfere – Ceolric charges him to suffer so just a cause – he is half won, half surprised – Offa, compelled to draw – they fight – Offa falls –

The soldiers are then ordered, reluctantly, by Envoy to seize him – he resists – and is wounded to Death.[1]

He supplicates Envoy, as he is a man and a soldier, to forbear punctilio as a Governour. He is overpowered.

[1] 'he says' is deleted.

Scene xii: El. 160, marked 5, provides a summary, which seems to offer two different possibilities:

Arnulph has received the fatal wound – as Elberta rushes in . . . he falls. Tenderly takes leave of her but exults in death that ends such a warpt existence.

Elberta enters crying out for her Children, frantic – seeing Arnulph recovers her –

The dialogue, which fits the first possibility, is from El. 157, 161, and 180. The following dialogue, from El. 167, dramatizes the second possibility:

Elberta (*entering wildly*)

Elberta. Where are my Children?
Are they all gone?
And died he with his Children?
O yes! they are dead! they murdered –
Tell me but one thing more –
Tell me the grave that holds their sweet remains
Charge them they look not on their wretched mother
Taint not their pure joy
With such a sight of woe –
Ah yes . . . I see them –
They mount the upper regions of the air
Ah See! – they beckon me! –

ELBERTA, ARNULPH.

ELBERTA. Hold! Stay! See! – 'tis his wife!
Her from whom every error had it's cause,
Her who or saves – or by his side expires –
ARNULPH. Cheaply, I sell my life – 'tis nothing worth –
Behold my bosom bare! But yet most dearly
I sell my Honour! – Huh! – my Honour, said I?
Proud fool! – tis lost! – No, 'tis my children's Honour,
My chaste wife's dignity I dearly sell.
I will not, by an ignominious end
Shame them in Death – howe'er in life I have blasted them.
Why what without them were prolong'd existence?
No! I have borne affliction as a man –
'Tis mortal destination –
But shame – disgrace – Avaunt! make way for Death!
Adieu, then, lov'd – lost Partner of my Heart!
O might the remnant of thy sorrowing Days
Pass in thy genial Home, with soften'd suffering
My little ones – adieu! Cling, cling not th‹us›

See you not that sweet smile?
How oft that charm'd me from
That loved, soft smile, that in this nether world
So oft has robb'd my Soul of[1] bitterness,
Chac'd from my cares their power
I come, my gentle loves! –
I see your soft arms opening to entwine
Your happy mother's neck – see those sweet cheeks
On which my lips have dwelt till time and spa‹ce›
And woe and all – but Arnulph – was forgot.
Arnulph? – O where is Arnulph?
Can you not tell me?
Is yonder his pale corse? –

A similar though briefer speech is found in El. 291:
ELBERTA. Tell me then, quickly – have you murder'd them?
Where are they? –
O do but lead me to their little graves
Let me but spread on them a few poor flow'rs.
[1] 'all its' is deleted.
V xii 1 s.d.–3 *ELBERTA, ARNULPH ... his side expires*: from El. 137, marked A5. The speaker's name is an editorial insertion.
V xii 4–10 *Cheaply ... blasted them*: from El. 161, marked A5.
V xii 10 *blasted them*: the following lines are added in pencil:
Tis only by this much I can redress them
This proof of wakened sensibility
To nature.
V xii 11 *Why*: before 'Why', 'What, what with' is deleted.
V xii 11–20 s.d. *Why what ... dying*: from El. 180, marked A5.

Around my tortur'd Heart –
Away! sad Images! Come, sterner Frost.

dying

Oh Death! to me thy pain
Is all unmixt with bitterness!
I'm now as blest to die – as, erst, to live.

Scene xiii

ARNULPH, ELBERTA, WILFRID, ACCA, TWO CHILDREN, CEOLRIC.

ARNULPH. Live, Live! O Live
To guard – protect – these little ones –

Dies.

ELBERTA. Ha! is he gone? Absolve me from my word
I will not live! –
WILFRID. 'Twas the last sound he ‹made.›
It seemed to soothe his exit. –

V xii 21–23 *Oh death ... erst, to live*: from El. 159.
Scene xiii: El. 156 and 154 provide a summary. The dialogue up to 'die! die! die!' is from El. 156; the remainder is from El. 154:

Arnulph is dying – when Wilfrid and Acca enter with the two Children – He just has voice to pronounce his blessing on them, and his charge that they obey and Cherish their mother – and to conjure her to live for their sakes – and expires.

Wilfrid says he has rescued them.

Informo brought in a Prisoner – offers to impeach.

Bid Peace! at this awful moment.

And too blasted to be saved by mean treachery since Offa already is slain – and Saxred fled.

Impeachment accepted to clear Ceolric –

Discovers the Casket to be spoil of Elberta's Father – now lawfully her own –

Elberta says why live ye now, ye little ones?

O die! die! die!

ELBERTA. WILFRID. ACCA. TWO CHILDREN. CEOLRIC.

ARNULPH dead.

Elberta feels too wretched for existence – she calls on her Children to weep for her but to weep no more for their Father – he is happy! and she must follow – Ceolric calls upon her to remember her promise to live for his sake – She asks if he attended to it? Yes, Wilfrid says, twas the last sound he heard, and seemed to compose his parting breath. She will hold it then, she says, sacred. They press her away – She begs to wait by his Corpse – but they force her off, accompanied by her weeping and affrighted Children.

Ceolric ends the Piece by a Moral on Strict Conduct Through all temptation and Trial of adversity.

ELBERTA. Then 'tis s‹acred.›
And I will live! – will torture out exist‹ence›
With more than murdering misery – My C‹hildren›
My little ones! – Why look ye there? – Come
Weep not for him – he's gone – 'tis true – but
He's gone to sweet forgiveness – O my Children
For me, for me your tender tears let f‹all›
He's gone where penitence is lost in
To pardon most benign – and tenderest mer‹cy.›
CEOLRIC. Did you not promise you would live?
ELBERTA. 'Tis true – and sacred be my plighted Word –
Yes! I will live! – to Heaven's high will resign'd –
I'm wondrous glad he's dead – for now I'm calm –
'Tis marvellous how I'm changed! I grieve at nothing.

V xiii 12 *lost in*: a word is torn away.
V xiii 1 s.d.–13 *ARNULPH, ELBERTA ... and tenderest mer‹cy›*: from El. 155, marked A5.
V xiii 14–16 *Did you not promise ... resign'd*: from El. 151. The speakers' names are editorial insertions.
V xiii 17–18 *I'm wondrous ... nothing*: from El. 153. The speaker's name is an editorial insertion.

Appendix

This appendix includes those fragments in which the characters' names indicate that the text belongs to *Elberta*, but which are either too general to place in a specific scene or which develop the plot in ways that conflict with the reconstructed text. Fragments that clearly belong to other plays and some that summarize the plot of *Elberta* in similar ways have not been included. See the textual note to Act I (p. 235 above) for the various plot summaries of *Elberta*.

1. From El. 87, marked A4. The speaker's name, an editorial insertion, comes from El. 193, which provides a one-line summary: 'She laments the perverse querulousness of wretchedness'. There is no place in Act IV for the speech, which resembles somewhat Elberta's soliloquy in Act I, Scene ix.

ELBERTA *sola*.

ELBERTA. Perverse oeconomy in wretchedness!
Where misery sharpest strikes her shaft envenom'd
We irritate the wound that calls for healing
By querulous dissonance, and fractious murmuring,
And all the goading pangs of sensibility
Urging impetuous feelings, quick suspicion,
Causeless resentment, and unkind rebuke.
O Patience! tenderest friend of worn calamity!

2. From El. 149 verso, cancelled. It may have been intended to form part of or to precede Act IV, Scene xvii, where Offa learns that the casket has not been delivered. There is no dialogue.

Eric informs that the flight of Elberta and the non appearance of Arnulph have been construed into guilt, and sentence is pronounced of confirmed confiscation of Elberta, in favour of Offa, and a new confiscation of Arnulph [an illegible word] himself, but not his family – from lenity.

3. From El. 184. The characters' names are editorial insertions (working backwards from the final speech), since Burney marks where new speeches begin but does not provide names. Elberta's last speech begins with the cancelled words 'Come to thy Children'. Cf. Act V, Scene ii.

ARNULPH. O Thou – to whom my Heart with ardent fondness –
ELBERTA. O spurn me! cast me off –
Ah! be not so unkind to pardon me!
ARNULPH. Resemble her in virtue – not in fate!
ELBERTA. O'erwhelming guilt
Despoil'd me of all energy; I sank . . .
ARNULPH. My poor Elberta led the way – but I –
O then – I could not follow –

4. From El. 194, marked A4. The speakers' names are editorial insertions. This might be the dialogue for Act IV, Scene ii, except that Arnulph's decision to return 'to arms and glory' suggests his later decision to fight Offa, not a decision to face trial, and Elberta does not argue with him in the summary. In Act V, he leaves without explaining why.

ELBERTA. And canst thou, Arnulph? – darst thou?
Hast thou no wife? no children?
ARNULPH. And then, to arms and glory I return;
Loaded with Fame to drive off ignominy,
Or, dying, cover'd fair with open wounds
My Heart's last blood shall blot out the foul stain.
O cruel Offa! well he knows I came
Unfurnish'd – void of means, or ways, or powers
Thee to redress –

5. From El. 195, marked A4. The speakers' names are editorial insertions. The dialogue resembles that of Act IV, Scene xv, but that scene concludes with Elberta's entry into the cave, terminating the dialogue.

ARNULPH. ELBERTA.

ARNULPH. In times distressful
Are there not moments when we lose the mastery
Of sober language? phrases unapt . . .
Absent – unmeaning – break in starts – abrupt –
'Tis not a mode of speech – but start of passion –
ELBERTA. O speak! explain –
Remove this dreadful chaos of surmise
That sinks by doubts – that awes by fearful mystery.
Involving every evil possible
In the wild space left for imagination
Midst the dense clouds of horrible obscurity!
O speak! explain!
That I at least may shape calamity!

6. From El. 243. The speakers' names are editorial insertions. It is unclear in which act this might belong.

ELBERTA. And all that we have liv'd on – the sole sustenance
Is the poor pittance Acca call'd her own,
And Wilfrid's bounty. To the neighbouring village
They hie alternate, and procure us food –
Food . . . for our little ones! –
Food – without which
Without these faithful, pitying, good old Servants
The high born babes of Arnulph and Elberta
Must perish! –
ARNULPH. Weepst then for this, my love? –
I feel exulting that Adversity
So stain'd with annals of neglect,
Ingratitude
Can thus in our sad tale shew beautiful!

7. From El. 248. Act I, Scene x has an interview between Elberta and Ceolric, but this dialogue fits neither the summary nor the existing dialogue. The characters' names were inserted later.

ELBERTA. CEOLRIC.

ELBERTA. I think . . . good General, you have a Friend!
CEOLRIC. I were a Monster else! for if a Man
I had not gained the whiteness of my brow
From chacing years that hunt me from the hills
Of blythsome youth, to Age's darken'd vale
Without some kindly act, some genial sympathy
That opening the pure Springs whose warmly currents
That urged me in my passage
As binds the honest Heart.

8. From Elberta 249. It is impossible to tell who is speaking.

Why art thou pale? Fearst thou the noble Arnulph
Afar though he head the Army of the Conqueror
His height in Rank, in Power, is poor, is lowly
To that which lifts him up above his fellows
In hardy valour, gentle courtesie
Excelling virtues, and exalting deeds.

9. From El. 254. The speakers' names are editorial insertions. Act V, Scene xi provides a confrontation between Offa and Arnulph, but the summary does not mention any accusations by Offa.

OFFA. What traitor's work to wed
  The Blood of my chief Foe?
ARNULPH. Traitor! – Ha! –
  If I had sought her
  If with deliberate scheme
  My Choice –
  But no Election gave her to my view –
  No project bad me gain her favour
  I saw her with the stern intent of capture.

10. From El. 265, marked Frag. Although this fragment has been thought by critics to provide 'the genesis of the play', its details conflict with the reconstructed plot. It is not even clear that it was intended for *Elberta*.

A Female is mentioned, who wild and unknown is seen roaming about – no one is informed whence she comes – woe is in her voice, terror in her aspect, – she never weeps, yet frequently wails, tho in terms unintelligible from their wildness –
  Her interesting appearance –
  Some she affrights – others is derided by –
  Her fierce harangues, though wild when offended –
  Her gentle supplications to shadows.
  Her inattention to pursuit –
  Her stated calls on.
  Stated visits to.

11. From El. 266 and verso. The speakers' names are editorial insertions. Offa seems to be attempting to force Arnulph to wed someone else in order to prevent his marrying Elberta. Cf. El. 54, in which Arnulph is said to be Offa's son:

Perhaps Arnulph *son* of Offa, occasions all the difficulty, from his hatred of Edgar Atheling and of Eric the last of his Chiefs and the Father of Elberta.

OFFA, ARNULPH.

OFFA. No more – no more –
  Wed her – or quit my presence and protection
ARNULPH. Hear me, then, Sir! – The Time is past! –
OFFA. Ha! past!

ARNULPH. O speak not yet! that frown of dire portent
OFFA. To whom? say quick to whom?
ARNULPH. Alas – I dare not
OFFA. Thou givst me a surmise of such fell sort
It strikes me to the Heart – my blood's swift current
Stagnates with poisonous swell –
O miscreant! if 'tis thus I have been deluded
Thus caught – trepann'd – ensnar'd – disgrac'd betray'd –
May all the darkest deities of revenge –
May the infernal Powers – Mayst thou endure –
Each torment that can scourge the Heart of man.
Then 'tis even so! – avaunt! – away for-ever!
O that some demon caught thee from my sight
To quicken my release! – cozen'd – abus'd –
Begone! – lest fury, mocking parentage,
Leave to the air vain words, and drive thee hence
By manlier means!

Appendix

The Triumphant Toadeater

A Musical Entertainment

In Two Acts

By Ralph Broome or Charlotte Burney, possibly in collaboration with Frances Burney

THE TRIUMPHANT TOADEATER

There is only one surviving manuscript of *The Triumphant Toadeater*, a fair copy of nine leaves (18 pages) in the Berg Collection. The handwriting is clear, though not neat, and there are no illegible words in the text. The various insertions and deletions, indicated in the textual notes, are all also legible, although some of the obliterations can be made out only with difficulty. The play has not been previously published.

In her *Companions without Vows* (1994), Betty Rizzo identifies, for the first time, the hand in which *The Triumphant Toadeater* is written as that of Ralph Broome, Charlotte Burney's second husband, whom she married on 1 March 1798. Rizzo also notes that the play 'is written on a high-grade paper with a distinctive watermark identical to paper he used for other writing in 1798'.[a] This identification leads Rizzo to attribute the play to Broome, at least as its probable author (p. 389). It is possible that Broome received aid in writing the play from Charlotte Burney, Frances's lively sister, or even from Frances herself. A few months after their marriage, the Broomes paid a visit to the d'Arblays; the play could have been written out at this time. In a letter of which the first six pages are missing, Burney notes that Broome was 'entertaining & informing', despite his arrogance; d'Arblay, however, 'too natural to like a character of this factitious sort, could not endure him' (*Journal and Letters*, IV, 174). Before his marriage to Charlotte, Burney had described him as a 'celebrated author', referring to his witty verse commentaries on the trial of Warren Hastings, *Letters of Simpkin the Second* (1789) (*Journals and Letters*, IV, 30).

Burney nowhere refers to *The Triumphant Toadeater* in her letters or journals, and Joyce Hemlow does not mention it in her biography or in her article on Burney's plays. Margaret Anne Doody believes that despite the play's Burneyesque title, 'the evidence is all against it' being hers:

> The mockery of the sexual feeling of an older woman is harsher than anything found in Burney's work, and none of Burney's constant themes (such as the value of work) are found. The language is coarser than Burney's in the plays and seems more based on dialect .... Sentence rhythms are also not typical of Burney. (*Frances Burney*, pp. 292, 419)

Doody believes that the play is 'evidently a piece by a woman, and probably by a Burney, or one of their circle'. Intriguingly, without having

[a] Rizzo, *Companions Without Vows: Relationships Among Eighteenth-Century British Women* (Athens: University of Georgia Press, 1994), p. 349, n. 38.

identified the hand as Ralph Broome's, she tentatively attributes the play to his wife Charlotte. Marjorie Lee Morrison terms *The Triumphant Toadeater* 'the slightest and least pretentious' of Burney's plays, but has no doubts about her authorship ('Fanny Burney and the Theatre', p. 174).

On her death, Burney's literary manuscripts were inherited by her niece Charlotte Barrett, Charlotte Burney's daughter. Barrett compiled a list of Burney manuscripts in her possession, of which number five is *The Triumphant Toadeater*.[a] The manuscript of the play bears the same number five, also in Barrett's hand. Barrett clearly assumed that the play was by Burney, rather than by her mother Charlotte Burney or her father Ralph Broome, but she was only two when the play was written and probably came across the play for the first time only after Frances Burney's death.

That *The Triumphant Toadeater* was written in summer 1798 is further suggested by the reference (p. 319) to the song 'Duet: Adieu' by Michael Kelly (1762–1826), an Irish actor, singer and composer who came to London in 1787 and was appointed Drury Lane's composer in 1797. The duet was first performed in the Haymarket Theatre's production of Isaac Bickerstaffe's comedy *Lionel and Clarissa* on 5 July 1798.[b] A copy survives in an undated collection of songs in the British Library (H.1652.ee (2)), set for two voices and piano, marked *andantino*, and entitled 'Adieu: a Favorite Duett as introduced in the Opera of Lionel & Clarissa':

Adieu adieu adieu my fears and sorrow,
Adieu adieu adieu my fears and sorrow,
Heav'n kindly mingled trouble,
Our tender joys to double,
Our tender joys to double,
That now in prospect lie.
Adieu adieu adieu my fears and sorrow,
Adieu adieu adieu my fears and sorrow.
Thus brighter shines the morrow,
When wint'ry tempests ending,
That beat all night descending,
The breath of morning fly,
The breath of morning fly –
Adieu adieu adieu my fears and sorrow,
Adieu adieu adieu my fears and sorrow,
Heav'n kindly mingled trouble,
Our tender joys to double,

[a] See Egerton ms. 3696, f. 173, in the British Library. The presence of *The Triumphant Toadeater* in this list is noted by Margaret M. Smith in *Index of Literary Manuscripts*, vol. III, part 1 (London: Mansell, 1986), p. 88. Smith wrongly, however, states that the list is in Burney's hand, when it is clearly in that of Charlotte Barrett. The manuscript bears a watermark of 1853, thirteen years after Burney's death, showing that Barrett compiled the list between then and her own death in 1870.

[b] See *The London Stage*, V, 2087.

Our tender joys to double,
That now in prospect lie.
Adieu adieu adieu adieu adieu.

By including the duet in the farce, the author could take advantage of the current popularity of Kelly's new song. There is nothing to indicate, however, that Broome (or his possible collaborators) ever attempted to have the play published or performed.

Men

Sir Nat Nonsuch	a poor Knight
Mr Mawkin	a Gentleman

Women

Mrs Torment	a widow Bewitched
Miss Backfriend	her Toadeater
Miss Roundabout	a Horsegodmother

Toadeater: a humble companion, dependent on another. The term, deriving from the practice of charlatans' attendants being employed to eat (or pretend to eat) toads, is applied to the heroines of *Evelina* (p. 294) and *The Wanderer* (p. 400).
Horsegodmother: a large, coarse-looking woman.

Scene a garden

Act I

Scene i

Enter Miss Roundabout *and* Miss Backfriend

Miss Roundabout. Mawkin returned? do you really think you saw Mr Mawkin in this wood yesterday?

Miss Backfriend. I'm sure I did. I have eyed him too often to be mistaken. Poor gentleman! so handsome as he is, it would be hard indeed if I did not remember his face though its so long since I have seen him. I have owed a grudge to that old hunks his father ever since he took him abroad to make him forget you. I'll answer for it the young Mawkin only thought of you the more out of spite, at least it is what I would have done myself in his place. And now I am persuaded that he is come over to England to carry you off nolens volens to Scotland.

Miss Roundabout (*laughing*). Oh dear! Miss Backfriend do you think so well it won't be nolens volens I promise you, for I shall be ready enough to go. I almost wish my Uncle Torment was at home, for he was quite on the Mawkin side of the question, to plague his Wife, and would not suffer her to lead me such a weary life about that little whipper snapper knight Sir Nat Nonsuch, who is deeply smitten with my fortune and still more with my prospects, for he has taken care to make sure that my aunt will leave every thing to me if I marry with her approbation. I received one of his odious epistles this morning which I put into the fire as usual unopened. But how came all these queer fancies into your head about Mr Mawkin and me? I never gave you any hint of our attachment.

Miss Backfriend. Oh! your aunt tells me every thing she knows about people; for indeed (ontry noos) she can keep nothing to herself. And then what with peeping at letters, listening at doors, and keeping constantly on the watch for what I can gather from winks and other intelligent signs, I get a great deal of useful information, which enables

I i 6 *hunks*: a surly, stingy old person.
I i 10 *nolens volens*: willing or unwilling; willy-nilly.
I i 10 *to Scotland*: i.e. to Gretna Green; see below, p. 316, and *The Witlings*, p. 70.
I i 24 *ontry noos*: i.e. *entre nous*; between ourselves.

I i 27 *which*: an insertion, replacing 'and'.

me to entertain my friends agreably, which makes my company always prefered to that of those dull and selfish mortals who are wrapped up in themselves and their own humdrum affairs.

MISS ROUNDABOUT (*aside*). What a string of impertinence! if this is the case the less you know of my affairs the better (*aloud*) really Miss Backfriend my aunt would be quite delighted to hear what pretty things you say of her ontry noos as you call it! and on one condition I promise not to inform her of it. And that is if you will prove that you *can* keep something to *yourself*, and don't mention Mr Mawkins return to England to her.

MISS BACKFRIEND. Oh! not for the world! if she knew it she would immediatly force you into a marriage with the threadbare Sir Nat, who wants her to fill his purse for him, and for that purpose is always paying his court to her. He would soon run out your fortune as he has done his own with his horses and dogs and nonsense I'll answer for it. But I hope you will take my advice, dissapoint them both and elope with Mr Mawkin. I will give you all the assistance in my power if you will go off with him some night.

MISS ROUNDABOUT. Yes – that would furnish you with agreable conversation for three months at least. But seriously Miss Backfriend you do not think I would be so wicked as to go to gretna green with Mr Mawkin, and marry him without my aunts consent. But say no more about it for I hear somebody coming.

MISS BACKFRIEND. Lest it should be Mrs Torment I'll make my escape for she is in a terrible humour to day. Adieu mon chêre remember Gretna green and command my services.

When it is too late to thwart you, the old Bugbear will forgive all depend upon it. And she cannot take your own fortune away do what you will.

Exit MISS BACKFRIEND.

Air MISS ROUNDABOUT.

MISS ROUNDABOUT.

I'll heave no more sighs, and I'll shed no more tears.
For all shall be gay, when my Mawkin appears.
Then haste lovely swain, and revive my sad heart.
For thy presence to Roundabout Joy will impart.

Exit MISS ROUNDABOUT.

I i 52 *Adieu mon chêre*: good-bye, my dear.

Scene ii

Enter Miss Roundabout, *followed by* Mr Mawkin.

Mr Mawkin. This meeting is fortunate indeed! most adorable of Women! and almost repays me for my inconceivable sufferings since our last melancholy parting. Knowing as I did that my father had jilted Mrs Torment I had not the courage to approach her house; but have wandered incessantly about these woods, since my return to England, in the faint hope of catching a distant glimpse of you, that I might seize the opportunity to die at your feet!

Yes – barbarous Miss Roundabout – your cruelty has broken a faithful heart which was intirely devoted to you. Behold the unhappy the injured Mawkin come to breathe forth his last sigh in your presence. Would that the happy Nonsuch had never been born, and that your heart still remained as unchanged as your person.

Miss Roundabout (*aside*). What beautiful language! I dare say that speech came out of some play (*aloud*) I have a rod in pickle for you Mr Mawkin. How dare you accuse me of barbarity, when I wrote you I don't know how many love letters which you never thought proper to answer. It is for me to wish that *your* heart remained as unchanged as *your* person Mr Mawkin.

Mr Mawkin. Charmer of my heart! you wring my soul! What is it you tell me? not receive my letters?

Miss Roundabout. Not one oh Mr Mawkin did you really write to me? into what perfidious hands have these letters fallen? what was their purport?

Mr Mawkin. They would have expressed to you my angel half the wretchedness I experienced in your abscence, how unable I was to bear so cruel a separation. They would have conveyed to you a faint idea of suffering too poignant to be fully described. The treacherous Torment has had the barbarity to conceal them we have been betrayed my charmer and cruelly decieved; read that letter and if you can help me to discover the author of it, I am now convinced that it came not from you.

Miss Roundabout *takes the letter and reads*

I ii 14 *a rod in pickle*: proverbial for a scolding or punishment in store.

I ii 21 *did you . . . me*: an insertion.
I ii 22 *these letters*: an insertion, replacing 'they'.
I ii 22 *was*: an insertion, replacing 'were'.
I ii 30 *am now . . . not*: an insertion, replacing 'know your hand too well to think for a moment that it came'.

MISS ROUNDABOUT.

To Mr Mawkin Esqr

Sir,

I am much surprized to find that you still persist in troubling me your letters contrary to the orders of Sir Mathew Mawkin and Mrs Torment. I know too well my duty to a beloved and respected aunt to be in any danger of following so bad an example even if I wished it. But rest assured sir that absence has cured me of the little love which I once foolishly felt for you, and that any future billy doos from you will be condemned to the flames.

Margaret Roundabout

P.S. Sir Nat Nonsuch is the happy man who will succeed you in my affections.

What unheard of impertinence! this writing is Miss Backfriends my aunts Toadeater. And billy doos is quite in her stile. I have no doubt about the authoress, and she shall feel my resentment. The young minx flatters herself probably that she will replace me in your affections, and aspires to the honor of becoming Mrs Mawkin.

MR MAWKIN. What! that low creature aspire to my hand! impossible! But do not throw all the blame on her. she is only the wretched tool of your inexorable aunt, who has no doubt dictated that letter. You see how it is my angel. My constancy to you has made me as hateful to your aunt as my father is from his inconstancy to her. I have now no hopes of gaining her consent to our union, and with many bitter tears resign myself patiently to my hard fate and bid you an eternal farewell – *weeps*.

MISS ROUNDABOUT (*aside*) What a mean coward! (*aloud*) Hold Mr Mawkin! do you think I will part with you thus? can you tamely submit to such base treachery, and quietly leave me to revenge it, while you sneak off in safety?

MR MAWKIN (*aside*). Bless me! I'm in a scrape. (*aloud*) What can I do my charmer? command your slave, would you have me fight a duel with Miss Backfriend?

MISS ROUNDABOUT (*sulkily*). I have no objection to it sir. but then my aunt would still remain unpunished.

MR MAWKIN. Perhaps you will wish me to try my skill again with her?

MISS ROUNDABOUT. I'm in no humour for joking Mr Mawkin. So tell me

I ii 37 *billy doos*: i.e. *billets-doux*; love-letters.

I ii 51 *father ... inconstancy*: an insertion, replacing 'fathers inconstancy made him'.
I ii 53 *patiently*: an insertion.
I ii 53 *an eternal*: an insertion.
I ii 54–5 *(aside) ... Mawkin!*: an insertion, replacing 'Hold Mr Mawkin! (*aside*) what a mean coward!'.
I ii 58 *Bless me*: an insertion, replacing 'Zounds'.
I ii 61 *(sulkily)*: an insertion.

instantly whether you will agree to take a little trip with me to Gretna green? that will be sport to us and death to them, and no duel in the question. Hey Mawkin what makes you look so down in the mouth?

MR MAWKIN. I confess I'm rather timid about venturing on such a wild scheme my angel suppose Mrs Torment should send Sir Nat Nonsuch after us?

MISS ROUNDABOUT. Paltry scruples! fear nothing; if *you* don't make him repent his insolence *I* will. I'll hear no more excuses, but will be here to meet you as soon as it is dark. Promise to be here.

MR MAWKIN (*aside*). I dare not refuse (*aloud*) most willingly, and thus I seal my promise. (*Kisses her hand*.) Till then farewell! most adorable of women.

and Exeunt.

Duett Adieu by Kelly.

Scene iii

Enter MRS TORMENT *and* MISS BACKFRIEND.

MRS TORMENT. That I'll be bound he is; but I would not advise him to shew his puppys face within my garden walls, for I shall make no scruple of giving him the second part of the same tune I gave his father old Sir Mathew when he was just such another stripling. He with nothing but his title to recommend him to jilt a young Lady of my fortune and attractions, who was a wife for one of the greatest Lords in the land, if they could but have had the wit to discover it. I wonder that my father ever proposed me to such a numbskul but that was all by my own desire to be sure for I thought of nothing but his title then and did not find out that he was such a Jackanapes. Well it shan't be my fault if there is not a Knight in the family however. And if they *do* marry I'll not leave the Jade a farthing of my money. Where did you see them together?

MISS BACKFRIEND. Here in this wood. I could not get near enough to pick up what they said, and only peeped occasionally from behind the trees at them. I was obliged at last to make off in haste as they kept a sharp look out to see if any one was near.

I ii 76 s.d. *Duett Adieu by Kelly*: Michael Kelly (1762–1826), an Irish actor, singer, and composer, wrote this duet for the production of Isaac Bickerstaffe's comedy *Lionel and Clarissa* at the Haymarket Theatre on 5 July 1798; see above, p. 312.

I ii 75–6 *most ... women*: an insertion.
I iii 6 *one of*: an insertion.
I iii 9 *thought of ... and*: an insertion.
I iii 10–11 *it shan't ... And*: an insertion

MRS TORMENT. I flattered myself that the letter we sent him would have made him despise her too thoroughly to suffer him to bear her again in his sight.

MISS BACKFRIEND. I fear he shewed her that letter, for I saw him give her one to read. But if he taxes me with it I'll deny it till I'm black in the face.

MRS TORMENT. Pray Miss Jenny give me leave to put in a word about my own affairs. It don't become you to be so meddling, though now I think of it I'm tired of the subject, and shan't worry myself any more about that unruly minx and her impertinent sweetheart. I've enough to do to think of my own lovers. They give me trouble enough one way or other. There's Mr Niggerly a sorry Knave has not been near me these six weeks.

MISS BACKFRIEND. No! to be sure ma'am, Miss Polecat will take care of that; she thinks she has him safe provided she keeps him out of your way. But she is no match for him for she has not wherewithal to keep a dog from starving. How many mean tricks she has played as I have heard from her maid to secure him. But it won't do depend it upon it ma'am you'll conquer at last.

MRS TORMENT. Well Jenny I *must* say that you're a cute creature and seem to know all things. I never met with your match for that. Pray can you tell me any thing about Captain Killall? for he has dropped even my correspondence?

MISS BACKFRIEND. Captain Killall! bless me did you never hear that story about his running away with Mrs Badenough?

MRS TORMENT (*angrily*). Where did you pick up such scandalous lies or when did you invent them Miss Chatterbox! Captain Killall would not dare play me such a trick knowing as he does the great regard I have for him. it is all your own invention to plague and mortify me. but I'll keep my temper in spite of you. Mr Torment never could put me out of temper in his life.

MISS BACKFRIEND (*aside*). Except for the day you boxed his ears and threw the inkstand at his head.

MRS TORMENT. What was that?

MISS BACKFRIEND. I said that I thought I saw Sir Nat Nonsuch coming Ma'am.

MRS TORMENT. I hope it is; for he's the only man in the world thats worth thinking of.

MISS BACKFRIEND. Except your husband you forgot to add that.

MRS TORMENT. True Jenny; but you know absence is a cure for love in

I iii 29 *Niggerly*: niggardly.
I iii 31 *Polecat*: a contemptuous term for a woman, often applied to a prostitute.

I iii 33 *she is no match*: the ms. reads 'she no match'.
I iii 41 *never hear that*: the ms. reads 'never that'.
I iii 44 *Chatterbox!*: 'Puppy as he is' is deleted.
I iii 52–3 *coming Ma'am*: an insertion.

all cases. But in *my* sad case Mr Torments presence had the same effect and was a most effectual cure for or rather a preventative to the tender passions. But talk no more about him I desire. But let me compose my countenance that I may be fit to be seen out of doors, I look so blowsy and have no glass to take a peep at myself in. What will he think of me Jenny?

MISS BACKFRIEND. He'll think you never looked better in your life ma'am, your walk (*aside* or your rouge) has given you a most becoming Colour!

Scene iv

MRS TORMENT, MISS BACKFRIEND, SIR NAT NONSUCH.

SIR NAT. My dear Mrs Torment I am overjoyed to see you out this fine day. You are really looking irresistible. Your servant Miss Backfriend – (*To* MRS TORMENT) may I enquire after the health of your lovely niece the charming Miss Roundabout?

MRS TORMENT. What can't you exist a moment without her? To what purpose then am I looking irresistible? oh Sir Nat I fear there are sad doings in that quarter; that whining Mawkin has taken advantage of Mr Torments absence to return and pester us with his love letters and visits. He durst not put his foot upon my threshold but that does not satisfy me, I must get you to kick him out of this neighbourhood Sir Nat.

SIR NAT. That I will with unfeigned pleasure the first time I meet with him. I'll shew him what metal I'm made of, and send his Jolterhead head to the devil or my name is not Nonsuch. Zounds! does he think I'll submit to be the rival of such a Maukish milksop! (*struts up and down the stage*.)

MRS TORMENT. Pray Sir Nat don't fluster yourself about them, I shall send for the child presently and she shall satisfy all your doubts. Here Jenny why don't you take the hint? go and fetch her immediately, but don't hurry her, as I have something particular to say to Sir Nat.

I iv 13 *Jolterhead*: a blockhead.
I iv 15 *Maukish*: mawkish, or sickly; playing on the name of 'Mawkin'.

I iii 59–60 *and was . . . passions*: an insertion.
I iv 1–2 *out this fine day*: an insertion.
I iv 2 *You are*: an insertion, replacing 'you're'.
I iv 2 *irresistible*: 'today' is deleted.
I iv 6 *looking*: an insertion.
I iv 9 *visits*: 'my house' is deleted.
I iv 18 *why don't you*: an insertion, replacing 'do you'.
I iv 19 *her*: an insertion.

MISS BACKFRIEND (*aside*). You want to have a têt o têt I see you old coquette, but I'll make it a short one. Mr Torment shall hear of all this before I die.

Exit MISS BACKFRIEND.

Scene v

MRS TORMENT *and* SIR NAT NONSUCH.

MRS TORMENT. Those were sweet lines you sent me Sir Nat. But you are a sad flatterer indeed. I have made my niece Roundabout set them to music, and she sings them prettily enough. The girl has a sweet pipe and some notion of music, I'll make her sing to you. But if you say they are your verses she will despise them immediately from the bottom of her soul: so have a care Sir Nat as you value her good opinion of your talents. And who knows but she may take it into her silly head to be jealous of me (*smirking*).

SIR NAT. Perhaps my too amiable friend there is more reason for jealousy than she is at present aware of (*takes* MRS TORMENT'S *hand*) Miss Roundabout would do well to hasten instead of retarding our nuptials. 'If she will not when she may, when she will she shall have nay.' for if I live to see you a widow – I'm a lost man.

Air

1

Torment of my life and Soul!
Compassionate the Pangs I feel
Thy dreaded frown inflicts a wound
Thy sweetest smile can never heal

2

But I alas! must bear the smart
My love conceal and hide my woe.
The hidden secret of my heart
Oh! Torment may'st thou never know.

MRS TORMENT (*aside, sighing bitterly*). Would I *was* free indeed! (*aloud*) My dear Sir Nat you are come too soon. My first husband Mr Torment still lives to plague me. Where he is, let those enquire who wish to know – as for me I have something else to think of than where

I iv 20 *têt o têt*: i.e. *tête à tête*; a private conversation between two persons.
I v 12 *If she will ... nay*: proverbial.

I v 11 *retarding*: an insertion, replacing 'retard'.

he's kicking up his heels. But as a letter he wrote to my niece lately give me reason to suspect that he is at present above ground, I must resign you Sir Nat. And exhort you to bear this trial with fortitude (*weeps*) and when you look at my Niece think of her once loved aunt, and give the poor girl a small portion of the love you have felt for me. and if fortune should ever be in our favor, if Mr Torment and his niece should one day leave us free, remember – there is one alive who does not hate you.

SIR NAT (*aside*). I must tie a knot in my handkerchief to remind me (*aloud*) oh! bitter moment! (*weeps*), (*aside*) What the devil shall I say to her? (*aloud*) yes dearest Madam, if the kind future should ever smile on us, think not that your charms will be neglected; some ten years hence perhaps –

MRS TORMENT (*in a passion*). Ten years! vile stripling base Jackanapes do you dare to laugh at me?

SIR NAT. Laugh Madam! witness these tears (*kneels and kisses her hand*)

MRS TORMENT. Rise and dry them quickly here come the children.

Scene vi

Enter MISS ROUNDABOUT, *and* MISS BACKFRIEND, *grinning*.

MRS TORMENT (*in confusion to* MISS ROUNDABOUT). I was Just giving Sir Nat my blessing my dear but I will take it from him again and give it to you if you behave as you ought, and do not disgrace me (my education of you I mean) for you're no more a relation of mine than Jenny Backfriend here (*knocking off* MISS BACKFRIEND'S *hat with her garden fan*). Mr Torment must bear the disgrace of you, if you are bent upon playing the fool with that stupid Mawkin whom you know I have the best reason in the world for hating. You have heard how his hair brained father Sir Matthew jilted me just as the young puppy means to do you. I did not think child that you would be so easily gulled. For my part (SIR NAT *sniggers*) I have been unfortunate 'tis true with regard to the tender passion, but always knew what I was about and who I had to deal with.

MISS ROUNDABOUT. Except Mr Niggerly aunt, Captain Killall, and Sir Mathew Mawkin, to say nothing of your husband. You have often told me that you were gulled by them all.

SIR NAT (*aside*). I could add another to that honorable list.

MRS TORMENT (*in a loud key*). I was never gulled in my life Miss

I v 38 *Jackanapes*: a ridiculous upstart.
I vi 10 *gulled*: duped; deceived.

I vi 5 *hat*: an insertion, replacing 'cap'.
I vi 11 *been unfortunate*: the ms. reads 'been been unfortunate'.

Impertinence; but I'll teach you to respect me before I've done with You, you great horse godmother. Go into the house, you are unworthy of the husband I intended for you, but you shall have him for all that, so come back and mend your manners. I'll have my own way at last. Such a child as you to choose a husband for herself indeed! I'm too good to let you have one at all. You deserve to die an old maid for your pertness like Jenny Backfriend there (*throws her glove at* MISS BACKFRIEND *who returns it to her*) who is pennyless except what she gets from me.

MISS BACKFRIEND (*weeping*).

Air

Oh! dear what will become of me
Dear dear what will become of me
Oh dear what will become of me
If I should die an old maid!
Have pity have pity upon a poor damsel sirs
Say did you ever hear such a hard case as hers
Leading for ever of nasty vile apes and curs
Oh! let her not die an old maid.

MRS TORMENT (*angrily*). If Miss Backfriend was to keep in the back ground, she would be more in her place, and I should be able to get in a word myself now and then in the course of a morning. Sir Nat I'm quite ashamed that you should see me in this taking, but these young Minxes put me almost beside myself, and make quite a different creature of me to what I am in general.

SIR NAT (*aside*). Your money's the same and that's enough for me. (*aloud*) Oh Madam! the gentleness of your nature is too well known for there to be any danger of my mistaking for a moment dignity and spirit, for ill humour and caprice. But (*in a whisper*) suppose you were to leave me with Miss Roundabout, and I'll answer for driving Mawkin out of her head, and placing Sir Nat Nonsuch there in his stead.

MRS TORMENT (*with fury*). What Sir, do you want to turn me out of my own garden? I protest I've never been so treated since the hour Mr Torment run away from me!

SIR NAT (*aside*). Hell and Tommy! help me out of this scrape. (*whispering* MRS TORMENT) After all that has passed between us my dear Madam, this supposition is barbarous.

I vi 27 *Air*: based on the anonymous eighteenth-century song 'Oh dear what can the matter be'.

I vi 34 *Leading ... vile apes*: from the proverbial 'old maids lead apes in hell'; see above, *Love and Fashion*, p. 174.

I vi 51 *Hell and Tommy*: the earliest example of this phrase in *OED* is from 1832–34: 'to play hell and tommy with', meaning to cause confusion.

I vi 38 *a morning*: an insertion, replacing 'an evening'.

I vi 40 *almost*: an insertion, replacing 'quite'.

MRS TORMENT (*aside to* SIR NAT). No Sir Nat, I cannot suspect you of indifference towards me. (*aloud*) Jenny come with me, Sir Nat Nonsuch has something very particular to say to my niece.

Exit MRS TORMENT *and* MISS BACKFRIEND.

MISS ROUNDABOUT. But as your Niece has nothing very particular to say to Sir Nat Nonsuch, she will lose no time in making a speedy retreat.

Exit MISS ROUNDABOUT.

SIR NAT.

Air

I ne'er a Horsegodmother saw in my life
So bearish, and so little formed for the Wife
    Of the sprightly ingenious Sir Nat
We shall never agree for a day or a night
For she'll henpeck me, scold me scratch quarrel and fight.
    We shall live in short like dog and cat.

Exit SIR NAT.

Act II

Scene i

Enter MISS ROUNDABOUT *followed by* SIR NAT NONSUCH.

MISS ROUNDABOUT. Here is this officious man again! And I'll again escape him if I can. *going*.

SIR NAT. Cruel fair one stay! grant me at least one moment to pour forth my soul at your feet.

MISS ROUNDABOUT. I am not disposed to give you 'my blessing' Sir Nat, so you will rise as unworthy as you kneel.

SIR NAT. Barbarous insinuation! is it possible that jealousy can be the cause of your coldness towards me? are you not satisfied that you possess my whole heart? must I renew those protestations of eternal Constancy which I have repeated so often?

MISS ROUNDABOUT (*drily*). You are at liberty to repeat them as often as you please sir, provided you do not expect me to stay here to listen. If you want an auditor I'll make Miss Backfriend come forward, for she is

I vi 58 *speedy*: an insertion, replacing 'timely'.

as usual upon the listening order behind one of the trees. I'll bring her to you presently – *going*.

SIR NAT (*detaining her*). Hear me Miss Roundabout! by thus trifling with my feelings you will drive me to desperation, neither Mrs Torment or her Toadeater have or ever shall have any place in my heart which here on my knees I dedicate to you for ever. (*Kisses her hand*).

MISS ROUNDABOUT. If you value your life Sir Nat fly this instant here is Mr Mawkin come to protect me

SIR NAT. Oh! the valiant Mawkin approaches does he? I'll shew him the length of my sword presently.

Scene ii

Enter MR MAWKIN

SIR NAT. Valiant Mawkin come on! defend your cause bad as it is. Now Roundabout for ever!

MR MAWKIN (*aside*). Lord help me! I'm in another scrape! my heart's in my breeches (*aloud*) – Come on Sir –

They fight. MAWKIN *pretends to be wounded and falls. He drops his sword which* MISS ROUNDABOUT *picks up and drives* SIR NAT *off the stage with it.*

SIR NAT. He's dead as a herring. Poor Mawkin! I'll make my escape out of the Kingdom.

Exit SIR NAT.

Scene iii

MISS ROUNDABOUT. Why how now Mawkin are you wounded man?

MR MAWKIN (*rising*). No thank god! that was a stratagem my dear creature to save my life. I feared it was all over with me.

MISS ROUNDABOUT. Bless me Mr Mawkin I am surprized at your want of spirit. Why I should beat you all to nothing in a battle.

MR MAWKIN. I hope you'll never try your skill upon me my dear. You are a brave girl truly, and if ever I'm drawn in to fight another duel you shall be my second. But now we have put our enemy to an ignominious

II ii 5 *dead as a herring*: proverbial.

II i 19 *for ever*: an insertion.

flight tell me quickly Goddess of my Idolatry, has mother Torment found out our little jaunt to Gretna green?

MISS ROUNDABOUT. Rest assured oh! valiant Mawkin that is a most profound secret. Though I confess that if your wonder working sword had not given me such timely assistance, I was on the point of confideing it to Sir Nat to get shot of him at once. We must break it to Mrs Torment soon, or I shall be married to the poor Knight nolens volens. For he is impatient to fill his purse with my money. But there is Miss Backfriend sneaking away, I'll make her come and explain the letter business.

runs off and returns in haste.

Oh Mawkin! it is my Aunt – hide yourself behind the trees.

Scene iv

Enter MRS TORMENT *in a passion.*

MRS TORMENT. Miss Backfriend *sneaking* indeed! she's no more given to sneaking than myself. Its very hard indeed if I cannot walk in my own grounds without being suspected of listening to your billing and cooing. But you have done for yourself Miss Roundabout. Every penny I have in the world shall go to Jenny Backfriend, my Toadeater in future I shall consider as my heir apparent. Ah you have reason to cry and sob, for I know all your tricks. A worthless couple you are as ever were linked together.

MISS ROUNDABOUT (*kneeling and weeping*). Yes Aunt I cannot deny it – Mawkin and I are one. Oh! pardon me this once.

MR MAWKIN (*kneeling*). Dear Mrs Torment give me your blessing.

MRS TORMENT (*knocking their heads together*). There is the only blessing you will ever recieve from me, so be thankful for it. (MR *and* MRS MAWKIN *squall*)

Scene v

MRS TORMENT MR *and* MRS MAWKIN SIR NAT NONSUCH *and* MISS BACKFRIEND. *The two latter run on and kneel to* MRS TORMENT.

MRS TORMENT. What another couple just arrived from Gretna Green? I

II iii 14 *shot of him*: rid of him.

II iii 10 *jaunt*: an insertion, replacing 'tour'.

had need be made of honey and sugar to bear all this. Do *you* want my blessing too Rabscallion?

SIR NAT. Dear Madam have patience and hear me with your usual sweetness. The lovely Miss Backfriend whom I understand from herself is now your heir apparent, has this moment offered me her hand; need I add that your approbation to our union alone is wanting to complete our happiness.

MRS TORMENT (*with emotion*). Rise Sir Nat! false and fickle, mean and avaricious as you are, rise. You have proved yourself unworthy even of my Toadeater, but I give her to you nevertheless. She shall become Lady Nonsuch and shall triumph over the headstrong Mrs Mawkin, who shall for ever rue the day when she refused you.

MISS ROUNDABOUT. Believe me aunt, when I have exposed Miss Backfriend's underhand conduct towards you and us all, you will find that she is backbiter general and less deserving of your favor than even the fickle knight.

MRS TORMENT. The more conspicuous their baseness, the greater my revenge on you and Mr Mawkin, who are a still more worthless couple. Such as they are they shall enjoy my fortune; for your disobedience has brought on you my eternal wrath. Jenny Backfriend shall replace you in my affections. But what courier so quickly carried you the news of your promotion Jenny?

MISS BACKFRIEND. I was taking the air in the wood ma'am and overheard all that passed between you and the Mawkins, and a great deal more which you shall be informed of anon.

MR MAWKIN (*to* MRS TORMENT). Will you suffer this insolence Madam?

MRS TORMENT. Suffer it Sir? I delight in it: it is quits to me to hear you affronted. The name of Mawkin has long been odious to me; you have completed its disgrace. I find sir that you inherit all your Fathers impertinence, it will be well for you if you inherit his fortune; for I again repeat that you have nothing to expect from me.

SIR NAT. Let me intreat you Madam not to renew this unpleasant subject, but let us dedicate this happy day to Joy and festivity. give me leave also to intercede for Mr and Mrs Mawkin. The former I had once the honor to kill.

MRS TORMENT. I will oblige you so far Sir Nat, as just to command my temper and forget my injuries to night. Though I cannot but think it hard that I am to sit down contented on this joyful Wedding day, with

II v 16 *backbiter general*: chief slanderer.

II v 7 *alone*: an insertion, replacing 'only'.
II v 12 *the*: an insertion, replacing 'my'.
II v 15 *underhand*: an insertion.
II v 15 *and us all*: an insertion.
II v 15–16 *that she is . . . and*: an insertion, replacing 'her'.
II v 27 *to* MRS TORMENT: an insertion.
II v 35–6 *The former . . . kill*: an insertion.

only my usual husband (and not even him now) and to go living on in the old humdrum way, to be sure nobody ever was so often crossed in love as I have been; I have always found that the more love I felt the less I inspired; and I dare say Mr Torment would have had the insolence to reject me if I had liked him. But I have not forgot my promise to you Sir Nat, I'll command my temper – through this day if these people will let me, as to night we will have a Ball, which I'll lead off with you. I have not danced these five and twenty years, and it will be an excellent practice for me. I will forgive even the two Mawkins for the present. (*she gives* MR *and* MRS MAWKIN *her hands to kiss*)

Air – A Fig for the treasures.

MR MAWKIN

Then let us be happy and chace away sorrow
Since we of forgiveness are only to taste.

MRS MAWKIN

And driving away all our cares till tomorrow
In grief let us not the time foolishly waste.

MR MAWKIN

What Torment can vex me now Roundabouts mine?

MRS MAWKIN

What Backfriend perplex me since Mawkin I'm thine?

Duett MR and MRS MAWKIN

Then aloud while we're blessing this thrice happy day
We'll not envy the Toadeaters triumphant sway

Grand Chorus.

But will all join our hands and our hearts while we may
And forgetting our quarrels be friends for to-day.

The Curtain Drops.

II v 49 s.d. *Air – A Fig for the treasures*: not identified.

II v 40 *in*: an insertion.